Human Activity Recognition and Behaviour Analysis

Liming Chen · Chris D. Nugent

Human Activity Recognition and Behaviour Analysis

For Cyber-Physical Systems in Smart Environments

Springer

Liming Chen
School of Computer Science
and Informatics
De Montfort University
Leicester, UK

Chris D. Nugent
School of Computing
Ulster University
Belfast, UK

ISBN 978-3-030-19407-9 ISBN 978-3-030-19408-6 (eBook)
https://doi.org/10.1007/978-3-030-19408-6

This Springer imprint is published by the registered company Springer Nature Switzerland AG.
The registered company address is: Gewerbestrasse 11, 6330 Cham, Switzerland

Preface

Recent advances in ubiquitous computing, sensing technologies, the Internet of Things, mobile computing and smart environments have led to a new wave of smart cyber-physical applications. These applications require human-machine systems to support context awareness, learning and cognitive capabilities, personalization and adaptation. Activity recognition and computational behaviour analysis are key to the success of such cyber-physical human-machine systems, as they provide essential contexts at multiple levels of abstraction for the aforementioned features. Over years, there has been a constant shift of sensor observation modelling, representation, interpretation and usage, from low-level raw observation data and their direct/hardwired usage, data aggregation and fusion to high-level formal context modelling, activity recognition and behaviour analysis, and further to change detection, semantic interpretation, decision support and recommendation. It is envisioned that this trend will continue towards a further higher level of abstraction, achieving situation, activity and goal awareness to support smart cyber-physical human-machine systems in a wide range of applications such as smart homes, smart grid, intelligent transport, smart cities, to name but a few.

This book provides a systematic introduction on human activity recognition and behaviour analysis for cyber-physical human-machine systems in smart environments. It is built upon the authors' extensive expertise, skills, experiences acquired over years' research, as well as insights and visions on future research trends and directions. The book focuses on knowledge-driven approaches and relevant underpinning technologies. It contains ten chapters, each targeting on a particular aspect. Chapters 1 and 2 set up the scene and context for the discussions of the remaining chapters of the book. Chapter 1 provides an overview of the background, basic concepts, existing approaches and methodologies, potential applications, opportunities and research trends and directions for computational behaviour analysis. Chapter 2 is dedicated to the comprehensive review on sensor-based approach to activity recognition, detailing the state of the art of both data-driven and knowledge-driven approaches as well as an in-depth critical analysis of their strengths and weaknesses. Chapters 3 and 4 introduce the ontology-based knowledge-driven approach to activity recognition. Chapter 3 focused on ontological activity modelling, sensor data analysis methods and semantic pattern reasoning mechanisms, whereas Chap. 4 details a hybrid approach, combining knowledge-driven activity recognition and data-driven behaviour analysis, to facilitating an incremental semi-automatic modelling process. Chapters 5 and 6 describe data segmentation approaches and methods for real-time streaming sensor data collected from a large number of miniaturised sensors and smart objects or more broadly the internet of things. Chapter 5 analyses the complexity and characteristics of activities of daily living, and introduces time window-based temporal segmentation methods and algorithms. Chapter 6 addresses the challenges of multiple activity recognition, namely interleaved and concurrent activities, by presenting an alternative, semantic-based approach to dynamic data segmentation. Chapter 7 introduces a solution to composite activity recognition, i.e. interleaved and concurrent activities, which combine ontological and temporal formalisms for activity modelling and reasoning in a multi-agent system. From Chap. 3 to Chap. 7, each chapter includes an example application case study, which uses real-world use scenarios and implemented system prototypes to test, evaluate and demonstrate the developed models, algorithms and methods. In addition to the underpinning approaches, methods and technologies described in previous chapters, Chaps. 8 and 9 describe the semantic smart home concepts and its underlying technologies and usage. It illustrates to readers how activity recognition and behaviour analysis are used in real-world problem solving. While Chap. 8 focuses on the technical creation of a semantic smart home, including the lifecycle of ontological modelling, semantic sensor data collection, storage and analysis, Chap. 9 concentrates on the use of semantic smart home for situation-aware assistance provisioning describing an agent-based assisted system for decision support and assistive living. Finally, Chap. 10 outlines four prototypes of user-centred cyber-physical systems covering a lightweight easy-to-deploy standalone system, a multi-agent system and two service-oriented system implementations based on different service technologies. This chapter is aimed at offering readers practical implementation knowledge and experience that they can easily adapt and use for their own research.

This book can serve many purposes. For undergraduate and Ph.D. researchers in relevant areas, it can be used as an introductory textbook to provide a comprehensive systematic introduction to activity recognition, covering the basic problem descriptions, concept definition, critical analysis of the methodologies and frameworks, state-of-the-art technologies, case studies and implementations, both theoretical and application aspects. For researchers such as postdocs, research associates or fellows, and junior academic, this book can be used as a handbook which provides in-depth descriptions for the knowledge-driven approach to activity recognition, covering the whole spectrum of the underlying technologies from modelling, data processing, pattern recognition, behaviour analysis to decision support and recommender systems. For academics and industrial practitioners in cyber-physical system domains, the book can be used as a practical reference book to provide a systematic methodology along with a scalable framework and extensible technology infrastructure, including models, methods, toolsets and prototype systems, which allow readers to follow up and build up by making best use of current research. For health and care technology and solution developers, this book offers an example technology solution for smart health care. All models, analysis methods, and technologies and systems are developed and implemented using the latest technologies in sensing, IoT, big data analysis, AI techniques and decision support, and tested and evaluated using real-world application scenarios from healthcare domain, e.g. assisted living and healthy ageing. It exemplifies a problem-solving process and a potential solution to many problems in smart health care.

Acknowledgements

The authors would like to express their gratitude to colleagues and research collaborators for their contributions to the content of this book. In particular, they would like to thank George Onyango Okeyo, Darpan Triboan, Joseph Rafferty and Ahmad Al-Bashrawi for their technical inputs in various research topics, implementation of prototype systems and evaluations of approaches and technologies in different application scenarios.

The authors would also like to thank Hui Wang, Roy Sterritt, Jesse Hoey, Diane J. Cook and Zhiwen Yu for their constructive discussions and comments during their collaborations in this research area.

The authors would also like to pay their heartfelt gratitude for the continued support over the years from their families for their encouragement and tolerance to the lost family engagements.

The authors would express their thanks to everyone in the Springer team, specifically Simon Rees, Wayne Wheeler and Manjula Saravanan, for their guidance and advice during this process.

Leicester, UK — Liming Chen
Belfast, UK — Chris D. Nugent

Contents

About the Authors

Liming Chen is Professor of Computer Science, Head of the Context, Intelligence and Interaction Research Group and its associated Smart Lab in the School of Computer Science and Informatics, De Montfort University, UK. He received his B.Eng. and M.Eng. from Beijing Institute of Technology, Beijing, China, and his Ph.D. in Artificial Intelligence from De Montfort University, UK. His research interests include data analytics, pervasive computing, user-centred intelligent systems, smart environments and their application in smart health and care. He is an IET Fellow, an IEEE Senior Member, a co-founder and co-director of the IEEE CIS "User-centred Smart Systems" Task Force. He has secured multi-million research funding, acting as coordinator and principal investigator, from the UK and European Union funding bodies, industry and third countries. He has over 200 peer-reviewed publications in internationally recognised journals and conferences, spanning both theoretical and applied research. He plays an active role in various scholarly activities, acting as a general or program chair for prestigious international conferences, and as editors and guest editors for a number of high-profile journals. He has delivered invited talks, keynotes and seminars in various forums, conferences, industry and academic events.

Chris D. Nugent is currently the Head of the School of Computing at Ulster University and leads the Pervasive Computing Research Group. He received a Bachelor of Engineering in Electronic Systems and D.Phil. in Biomedical Engineering both from Ulster University and currently holds the position of Professor of Biomedical Engineering. His research within biomedical engineering addresses the themes of the development and evaluation of technologies to support pervasive healthcare within smart environments. Specifically, this has involved research in the topics of mobile based reminding solutions, activity recognition and behaviour modelling and more recently technology adoption modelling. He has published over 450 papers in these areas and has been a grant holder of Research Projects funded by National, European and International funding bodies. He is the co-Principal Investigator of the Connected Health Innovation Centre at Ulster University and a co-Investigator of the British Telecom Ireland Innovation Centre. In 2016 he was awarded the Senior Distinguished Research Fellowship from Ulster University.

Abbreviations

AAL	Ambient Assisted Living
A-Box	Assertion Box
ADL	Activities of Daily Living
AJAX	Asynchronous JavaScript and XML
API	Application Programming Interface
AR	Activity Recognition
BLE	Bluetooth
CEP	Complex Event Processing
CRUD	Create, Read, Update and Delete
DD	Data-driven
EC	Event Calculus
ESB	Enterprise Service Bus
HAR	Human Activity Recognition
HCI	Human–Computer Interface
HTML	Hypertext Markup Language
HTTP	Hypertext Transfer Protocol
ICT	Information and Communications Technology
IoT	Internet of Things
JESS	Java Expert System Shell
JSON	JavaScript Object Notation
KD	Knowledge-driven
MQTT	Message Queuing Telemetry Transport
OS	Operating System
RDBMS	Relational Database Management System
RDF	Resource Description Framework
REST	Representational State Transfer
RF	Radio Frequency
SH	Smart Home
SOA	Service-oriented Architecture
SOAP	Simple Object Access Protocol

SPARQL	SPARQL Protocol and RDF Query Language
SWRL	Semantic Web Rule Language
T-Box	Terminology Box
TDB	Triplestore Database
USB	Universal Serial Bus
XML	Extensible Markup Language

List of Figures

List of Tables

Chapter 1
Introduction

1.1 Background

Recent advances in sensing technologies, the Internet of Things (IoT), pervasive computing, smart environments have transformed traditional embedded Information and Communications Technology (ICT) systems into an ecosystem of interconnected and collaborating smart objects, devices, embedded systems and most importantly humans. Such systems, often referred to as cyber-physical systems (CPS), are usually human user-driven or user–centred, and are aimed at providing people and businesses with a wide range of innovative applications and services, e.g. making smarter, more intelligent, more energy-efficient and more comfortable our transport systems, cars, factories, hospitals, offices, homes, cities and personal devices. For example, a "Smart Home" can monitor and analyse the daily activities of its inhabitants, usually the elderly or individuals with disabilities, so that personalised context-aware assistance can be provided. A "Smart City" can monitor, manage and control all basic city functionalities such as transport, energy supply and waste collection, at a higher level of automation by collecting and harnessing sensor data across the larger geographic expanse.

In order to respond in real-time to an individual user's specific needs in dynamic and complex situations and support ergonomic and user-friendliness taking into various human factors such as privacy, dignity and behaviour characteristics, cyber-physical human-machine systems need to be aware of the physical environment and human participants' behaviour. This awareness enables effective and fast feedback loops between sensing and actuation, possibly with cognitive and learning capabilities adapting to the participants' preferences, capabilities and the modality of human-machine interaction as well as dynamic situations. At this moment, vigorous research on cyber-physical human-machine systems and their applications have been undertaken in various national, regional and international research initiatives and programs. These include the smart home for supporting active and healthy ageing, and smart cities to enhance performance and wellbeing, to reduce costs and resource consumption, and to engage more effectively and actively with its citizens. A whole

L. Chen and C. D. Nugent, *Human Activity Recognition and Behaviour Analysis*, https://doi.org/10.1007/978-3-030-19408-6_1

raft of heterogeneous computing technologies providing fragments of the necessary functionality, e.g. sensor networks, data analytics, artificial intelligence, pervasive computing, human-computer interaction, has been developed. Nevertheless, to support the core features of this new breed of smart cyber-physical applications with context-awareness, personalisation and adaptation, computational agents, devices and the overall human-machine systems should be activity and goal aware as well as responsive to changes. The collection, modelling, representation and interpretation of the states, events and behaviours happened between human participants and both physical and software agents/devices situated in a cyber-physical integrated environment need to be carried out in a formal systematic way and at a higher level of abstraction. This will facilitate data fusion and joint interpretation based on multiple dimensions of observations (e.g., environment context, human physical activities and mental or system states) as well as longitudinal pattern recognition.

Over the past decade, the modelling, representation, interpretation and usage of sensor observations have constantly shifted from low-level raw observation data and their direct/hardwired usage, data aggregation and fusion, to high-level formal context modelling and context-based computing. It is envisioned that this trend will continue towards a further higher level of abstraction, allowing situation, activity and goal modelling, representation and inference. The resulting technologies will thus enable and support user-centred functionality, ergonomics and usability in the next generation of smart cyber-physical applications.

Human activity recognition (HAR) is key to the successful realisation of intelligent cyber-physical systems. This relates to the fact that activities in a pervasive environment provide important contextual information and any intelligent behaviour of situated CPSs within the environment must be relevant to the user's context and ongoing activities. As such, HAR has become one of the most important research topics for a variety of problem domains, including pervasive and mobile computing [1], surveillance-based security [2], context-aware computing [3] and ambient assistive living [4]. With the prevalence of the underlying technologies, e.g. IoT, data analytics, artificial intelligence, HCI and pervasive computing, and the acceptance and industry-level uptake of the new wave of CPS applications in smart healthcare, smart cities, intelligent transport and energy, it becomes increasingly apparent that activity recognition and computational behaviour analysis will play a decisive role in the future smart CPSs.

1.2 Basic Concepts on Activity Recognition

1.2.1 Action and Activity

Before we embark on an in-depth discussion on activity monitoring, modelling and recognition, it is useful to distinguish human behaviours at different levels of granularity. For physical behaviours, the terms "action" and "activity" are commonly

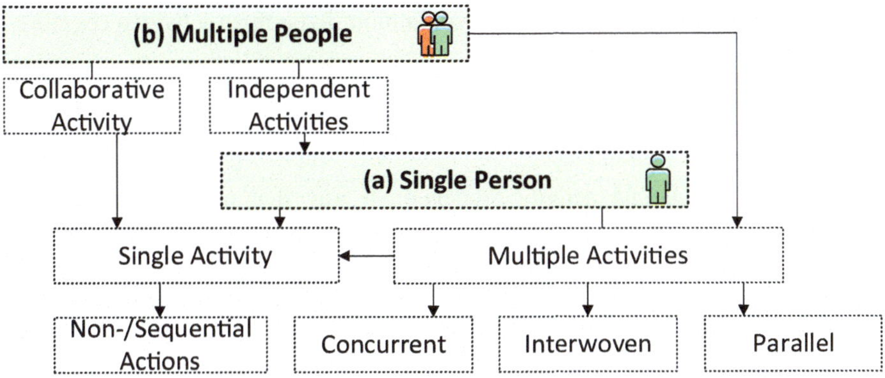

Fig. 1.1 Activity classification: **a** single-user composite activity, **b** multiple user composite activity

used in activity recognition communities. In some cases, they are used interchangeably and in other cases they are used to denote behaviours of different complexity and duration. In the latter cases the term "action" is usually referred to as simple ambulatory behaviour executed by a single person and typically lasting for a very short duration of time. Examples of actions include bending, retrieving a cup from a cupboard, opening a door, putting a teabag into a cup, etc. On the other hand, the term "activity" here refers to complex behaviours consisting of a sequence of actions and/or interleaving or overlapping actions. They could be performed by a single human or several humans who are required to interact with each other in a constrained manner. They are typically characterized by much longer temporal durations, such as making tea or two persons making meals. As one activity can contain only one action, there will be no cut-off boundary between these two behaviour categories. Nevertheless, this simple categorization provides a baseline on the concept of action and activity for the discussions in this book.

Activities can be performed in many contexts. A single person can perform a single activity sequentially or multiple activities in a sequential, concurrent, interwoven or parallel manner. On the other hand, multiple people can perform a single activity or multiple activities in a sequential, concurrent, interwoven or parallel manner in a shared space, independently or in collaboration. Figure 1.1 show a rough hierarchical structure for categorising human activities in terms of the numbers of involved users and activities.

1.2.2 Activity Recognition

Activity recognition is the process whereby a person's behaviour and his/her situated environment are monitored and analysed to infer the undergoing activities. It comprises many different tasks, namely activity modelling, behaviour and environment monitoring, data processing and pattern recognition. To perform activity recognition, it is, therefore, necessary to:

(1) choose and deploy appropriate sensors to objects and environments in order to monitor and capture a user's behaviour along with the state change of the environment;
(2) create computational activity models in a way that allows software systems/agents to conduct reasoning and manipulation;
(3) collect, manage and process perceived information through aggregation and fusion to generate a high-level abstraction of context or situation;
(4) design and develop reasoning algorithms to infer activities from collected sensor data;
(5) carry out pattern recognition to determine the performed activity.

Researchers from different application domains have investigated activity recognition for the past two decades by developing a diversity of approaches and techniques for each of these core tasks. It is often the case that the selection of a method used for one task is dependent on the method of another task(s), which will be closely examined and studied in the remaining of this book.

These concepts about action, activity, the way of performing activities and the characterisation of activity recognition tasks discussed above provide a basic conceptualisation and clarity for the discussions in this chapter and the rest of this book.

1.3 Activity Recognition Approaches

Monitoring an actor's behaviour along with changes in the environment is a critical task in activity recognition. This monitoring process is responsible for capturing relevant contextual information for activity recognition systems to infer an actor's activity. In terms of the way and data type of these monitoring facilities, there are currently two main activity recognition approaches; vision-based activity recognition and sensor-based activity recognition.

1.3.1 Vision-Based Activity Recognition

Vision-based activity recognition uses visual sensing facilities, e.g., camera-based surveillance systems, to monitor a person's behaviour and its environment changes. The generated sensor data are video sequences or digitized visual data. The approaches in this category exploit computer vision techniques, including feature extraction, structural modelling, movement segmentation, action extraction and movement tracking to analyse visual observations for pattern recognition. Vision-based activity recognition has been a research focus for a long period of time due to its important role in areas such as human-computer interaction, user interface design, robot learning and surveillance. Researchers have used a wide variety of modalities, such as single camera, stereo and infrared, to capture activity contexts. In addition, they have investigated a number of application scenarios, e.g., single actor or group tracking and recognition.

Vision-based activity recognition has been a research focus for a long period of time due to its important role in areas such as surveillance, robot learning and anti-terrorist security. Researchers have used a wide variety of modalities, such as a single or multi-camera, stereo and infra-red, to investigate a diversity of application scenarios, for single or groups of individuals. Several survey papers about vision-based activity recognition have been published over the years. Aggarwal and Cai [5] discuss the three important sub-problems of an action recognition system extraction of human body structure from images, tracking across frames, and action recognition. Cedras and Shah [6] present a survey on motion-based approaches to recognition as opposed to structure-based approaches. Gavrila [7] and Poppe [8] present surveys mainly on tracking human movement via 2D or 3D models and the enabled action recognition techniques. Moeslund et al. [9] presents a survey of problems and approaches in human motion capture, tracking, pose estimation, and activity recognition. Yilmaz et al. [10] and Weinland et al. [11] present surveys of tracking objects for action recognition. More recently, Turaga et al. [2] and Aggarwal et al. [12] present surveys focusing on high-level representation of complex activities and corresponding recognition techniques. Together these works have provided an extensive overview on the vision-based approach. Given these existing works, this chapter will not review research on vision-based activity recognition. However, it is worth pointing out that while visual monitoring is intuitive and information-rich, vision-based activity recognition suffers from issues relating to privacy and ethics [10] as cameras are generally perceived as recording devices. Compared with the number of surveys in vision-based activity recognition and considering the wealth of literature in sensor-based activity recognition, there is a lack of extensive review on the state of the art of sensor-based activity recognition. This may be because the approach only, recently, became feasible when the sensing technologies matured to be realistically deployable in terms of the underpinning communication infrastructure, costs, and sizes.

It is worth pointing out that while visual monitoring is intuitive, information-rich, and considerable work has been undertaken and significant progress has been made, vision-based activity recognition approaches suffer from issues related to scalability and reusability due to the complexity of real-world settings, e.g., highly varied activities in the natural environment. In addition, as cameras are generally used as recording devices, the invasiveness of this approach as well as privacy and ethics concerns as perceived by some also prevent it from large-scale uptake in some applications, in particular, in home environments.

1.3.2 Sensor-Based Activity Recognition

Sensor-based activity recognition is to use the emerging sensor network technologies, the Internet of Things (IoT), and smart devices for activity monitoring. The generated sensor data from sensor-based monitoring are mainly time series of state changes and/or various parameter values that are usually processed through data fusion, probabilistic or statistical analysis methods and formal knowledge technologies for activity recognition. A wide range of sensors, including contact sensors, RFID, accelerometers, audio and motion detectors, to name but a few, are available for activity monitoring. These sensors are different in types, purposes, output signals, underpinning theoretical principles and technical infrastructure. Sensors can be attached to either an actor under observation or objects that constitute the environment. Sensors attached to humans, i.e., wearable sensors, often use inertial measurement units (e.g. accelerometers, gyroscopes, magnetometers), vital sign processing devices (heart rate, temperature) and RFID tags to gather a person's behavioural information. Activity recognition based on wearable sensors has been extensively used in the recognition of human physical movements characterised by a distinct, periodic motion pattern such as physical exercises, walking, running, sitting down/up, climbing.

The wearable sensor-based approach is effective and relatively inexpensive for data acquisition and activity recognition for certain types of human activities, mainly human physical movements. Nevertheless, it suffers from two drawbacks. Firstly, most wearable sensors are not applicable in real-world application scenarios due to technical issues such as size, ease of use and battery life in conjunction with the general issue of acceptability or willingness of the user to wear them. Secondly, many activities in real-world situations involve complex physical motions and complex interactions with the environment. Sensor observations from wearable sensors alone may not be able to differentiate activities involving simple physical movements, e.g., making tea and making coffee. To address these issues, object-based activity recognition has emerged as one mainstream approach. The approach is based on real-world observations that activities are characterised by the objects that are manipulated during their operations. Simple sensors can often provide powerful clues about the activity being undertaken. As such, it is assumed that activities can be recognised from sensor data that monitor human interactions with objects in the environment.

Sensors attached to an object within an environment, namely ambient sensors are used to infer activities by monitoring human-object interactions through the usage of multiple multi-modal miniaturized sensors. As such, this approach is often referred to as dense sensing as it involves in the use of many ambient sensors. It is particularly suitable for dealing with activities that involve a number of objects within an environment. Research on this approach has been driven by the intensive research interest and huge research effort on smart home based assistive living, such as the EU's Active Assisted Living (AAL) program. In particular, sensor-based activity recognition can better address sensitive issues in assisted living such as privacy, ethics and obtrusiveness than conventional vision-based approaches. This combination of application needs, and technological advantages has stimulated considerable research activities in a global scale, which gave rise to a large number of research projects, including the House_n [13], CASAS [26], Gator-Tech [14], inHaus [15], AwareHome [16], DOMUS [17] and iDorm [18] projects, to name but a few. As a result of the wave of intensive investigation, there have seen a plethora of impressive works on sensor-based activity recognition in the past several years [19, 20].

Object-based activity recognition has attracted increasing attention as low-cost low-power intelligent sensors, wireless communication networks and pervasive computing infrastructures become technically mature and financially affordable. It has been, in particular, under vigorous investigation in the creation of intelligent pervasive environments for ambient assisted living (AAL), i.e., the SH paradigm. Sensors in a SH can monitor an inhabitant's movements and environmental events so that assistive agents can infer the undergoing activities based on the sensor observations, thus providing just-in-time context-aware ADL assistance. For instance, a switch sensor in the bed can strongly suggest sleeping, and pressure mat sensors can be used for tracking the movement and position of people within the environment.

It is worth pointing out that the approaches described above may be suitable for different applications because the sensing devices come in different in size, weights, cost, measurement, mechanism, software, and communication and battery life. Taking this into account it is not possible to claim that one approach is superior to the other. The suitability and performance in the end, is down to the nature of the type of activities being assessed and the characteristics of the concrete applications. In most cases, they are complementary and can be used in combination in order to yield optimal recognition results.

1.4 Activity Recognition Methods

Activity recognition methods can be broadly divided into two major categories. The first one is based on data mining and machine learning techniques while the second strand of methods is based on priori domain knowledge and logical modelling and reasoning. The former is usually referred to as data-driven approach, and the latter as knowledge-driven approach. Both are elaborated below.

1.4.1 Data-Driven Activity Recognition

Data-driven methods for activity recognition include supervised and unsupervised learning methods, which primarily use probabilistic and statistical reasoning. Supervised learning requires the use of labelled data upon which an algorithm is trained. Following training the algorithm is then able to classify unknown data. The general procedure using a supervised learning algorithm for activity recognition includes several steps, namely, (1) to acquire sensor data representative of activities, including labelled annotations of what an actor does and when, (2) to determine the input data features and its representation, (3) to aggregate data from multiple data sources and transform them into the application-dependent features, e.g., through data fusion, noise elimination, dimension reduction and data normalization, (4) to divide the data into a training set and a test set, (5) to train the recognition algorithm on the training set, (6) to test the classification performance of the trained algorithm on the test set, and finally (7) to apply the algorithm in the context of activity recognition. It is common to repeat steps (4)–(7) with different partitioning of the training and test sets in order to achieve better generalisation with the recognition models. There are a wide range of algorithms and models for supervised learning and activity recognition. These include Hidden Markov Models (HMMs), dynamic and naive Bayes networks, decision trees, nearest neighbour and support vector machines (SVMs) [21, 22]. Among them HMMs and Bayes networks are the most commonly used methods in activity recognition.

Unsupervised learning on the other hand tries to directly construct recognition models from unlabelled data. The basic idea is to manually assign a probability to each possible activity and to predefine a stochastic model that can update these likelihoods according to new observations and to the known state of the system. Such an approach employs density estimation methods, i.e., to estimate the properties of the underlying probability density or clustering techniques, to discover groups of similar examples to create learning models. The general procedure for unsupervised learning typically includes (1) to acquire unlabelled sensor data, (2) to aggregate and transforming the sensor data into features, and (3) to model the data using either density estimation or clustering methods. Algorithms for unsupervised learning include the use of graphical models [23] and multiple eigenspaces [24]. A number of unsupervised learning methods are also based on probabilistic reasoning such as various variants of HMMs and Bayes networks. The main difference between unsupervised and supervised probabilistic techniques is that, instead of using a pre-established stochastic model to update the activity likelihood, supervised learning algorithms keep a trace of their previous observed experiences and use them to dynamically learn the parameters of the stochastic activity models. This enables them to create a predictive model based on the observed agent's activity profiles.

A major strength of the activity recognition algorithms that are based on probabilistic learning models is that they are capable of handling noisy, uncertain and incomplete sensor data. Probabilities can be used to model uncertainty and capture domain heuristics, e.g., some activities are more likely than others. The limitation of the unsupervised learning probabilistic methods lies in the assignment of these handcrafted probabilistic parameters for the computation of the activity likelihood. They are usually static and highly activity-dependent. The disadvantage of supervised learning in the case of probabilistic methods is that they require a large amount of labelled training and test data. In addition, to learn each activity in a probabilistic model for a large diversity of activities in real-world application scenarios could be deemed as being computationally expensive. The resulting models are often ad hoc, not reusable and scalable due to the variation of the individual's behaviour and their environments.

1.4.2 Knowledge-Driven Activity Recognition

Knowledge-driven methods for activity recognition is to exploit knowledge modelling and representation for activity and sensor data modelling, and to use logical reasoning to perform activity recognition. The general procedure of a knowledge-driven approach includes (1) to use a knowledge representation formalism to explicitly define and describe a library of activity models for all possible activities in a domain; (2) to aggregate and transform sensor data into logical terms and formula; and (3) to perform logical reasoning, e.g., deduction, abduction and subsumption, to extract a minimal set of covering models of interpretation from the activity model library based on a set of observed actions, which could explain the observations.

There exist a number of knowledge modelling and representation methods and reasoning algorithms in terms of logical theories and representation formalisms. For example, Kauz [25] adopted first order axioms to build a library of hierarchical plans for plan recognition. Wobke [26] extended Kauz's work using situation theory to address the different probabilities of inferred plans. Bouchard [27] used action Description Logic (DL) and lattice theory for plan recognition with particular emphasis on the modelling and reasoning of plan intra-dependencies. Chen [28] exploited the event theory—a logical formalism, for explicit specification, manipulation and reasoning of events, to formalise an activity domain for activity recognition and assistance. The major strength of Chen's work is its capabilities to handle temporal issues and undecidability. Logical activity modelling and reasoning is semantically clear and elegant in computational reasoning. It is also relatively easy to incorporate domain knowledge and heuristics for activity models and data fusion. The weakness of logical approaches is their inability or inherent infeasibility to represent fuzziness and uncertainty.

Most of them offer no mechanism for deciding whether one particular model is more effective than another, as long as both of them can be consistent enough to explain the actions observed. There is also a lack of learning ability associated with logic-based methods.

1.5 Activity Recognition Applications

From an application perspective, activity recognition is seldom the final goal but usually one step of an application system. For example, in assistive living, activity recognition is used as input for decision making that attempts to detect and provide activity assistance. In security applications, activity recognition helps identify potential troublemakers providing an input for the following investigation and decision-making processes. While it is beyond the scope of the chapter to provide a thorough review on activity recognition applications, Table 1.1 summarises the major application categories and some key application areas for reference.

Table 1.1 The application categories and example areas

Application categories	Example application areas
Security	Airport, train station, Banks, car park warning systems
Intelligent environment	Smart homes, smart offices, meeting rooms, smart hospitals, smart cars, smart classrooms
Healthcare	Activity (physical) tracking, assistive living, well-being monitoring
Military	Soldier monitoring in the battlefield
Pervasive and mobile computing	Context-aware interface design, content delivery, mobile task assistance, energy efficiency

1.5.1 A Typical Application Scenario: Ambient Assisted Living

Ambient-assisted living (AAL) aims to exploit activity monitoring, recognition, and assistance to support independent living and ageing in place. Other emerging applications, such as intelligent meeting rooms and smart hospitals, are also dependent on activity recognition in order to provide multimodal interactions, proactive service provision, and context-aware personalised activity assistance. The main goal of building an AAL system is to aid the inhabitants in a Smart Home (SH) environment to carry out their Activities of Daily Living (ADL). A SH is considered to be augmented living environments equipped with sensors and actuators, within which monitoring of ADL and personalised assistance can be facilitated. Several lab-based or SH systems have been developed to support inhabitants in conducting ADLs such kitchen-based activities, taking medications, detecting anomalies and behaviour patterns. However, these SH systems provide only the fragments of necessary functionality required to support for independent living. The existing SH technologies and solutions suffer from a number of drawbacks. One of the key limitations is the lack of interoperability of the systems between vendors, adapting non-standard communications protocols and developing proprietary devices. This creates a huge challenge in reusability of the SH system for not only integrating third-party sensors and actuators devices but also limit heterogeneity of the data. Hence, the adaptation and applicability of solutions affect the end-user.

The fundamental processes undertaken by the AAL system can be categorised in three Ps: Preparing, Processing and Presenting [29]. The preparing stage involves developing activity models, data collection and monitoring. The processing stage comprises of segmenting the raw data stream, inferencing and recognising mixed user (also referred to inhabitant) activities, aiding when required and learning new activities. The resource-intensive processing tasks are generally delegated from resource-constrained devices to more powerful devices such as servers with a web service interface. The presenting stage is responsible for tailoring the system to specific application types and providing an intuitive human-computer interface (HCI). Figure 1.2 illustrates these phases as the building blocks of an AAL system.

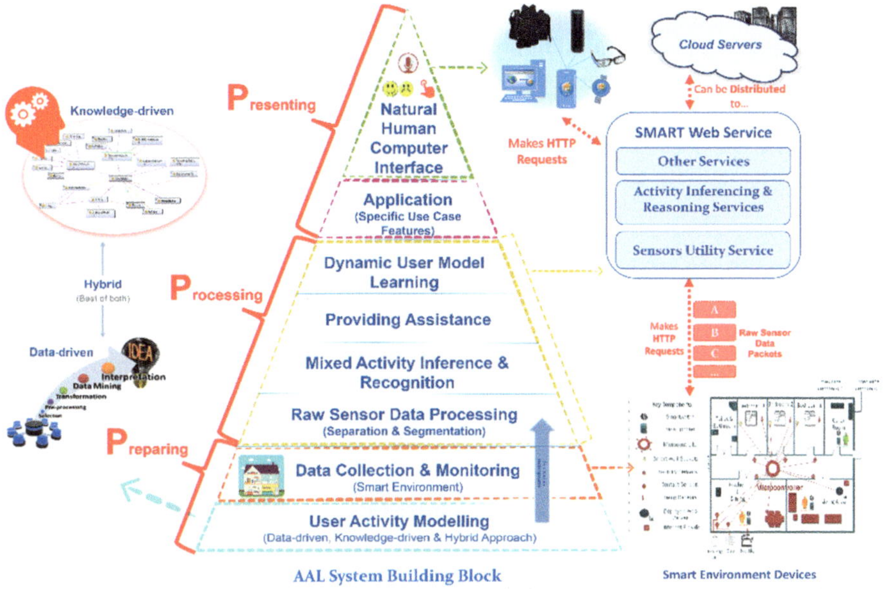

Fig. 1.2 The 3-Ps AAL framework: preparing, processing and presenting

1.5.2 Activity Recognition Challenges in Ambient Assisted Living

Activity recognition in the context of ambient assisted living within a smart home presents a number of challenges. Firstly, ADLs can be carried out with a high degree of freedom in relation to the way and the sequential order they are performed. Individuals have different lifestyles, habits or abilities and as such have their own way of performing ADLs. Though ADLs usually follow some kind of pattern there are no strict constraints on the sequence and duration of the actions. For example, to prepare a meal one can firstly turn on the cooker and then place a saucepan on the cooker, or vice versa. Such phenomena happen in almost all ADLs, e.g., preparing a drink, grooming, to name but a few. The wide range of ADLs and the variability and flexibility in the manner in which they can be performed require an approach that is scalable to large scale activity modelling and recognition.

Secondly, multi-modal sensors co-exist in an SH. They generate heterogeneous data different in both formats and semantics. It is often necessary to fuse and interpret sensor data from multiple sources in order to establish the context of the ongoing ADL. For instance, the ADL making tea may involve the preparation of tea bag, cup, hot water, milk and sugar. Only when some or all sensor data from these items are fused, can the ADL be recognised. In addition, sensor data are full of noises, e.g., missed activations and/or faulty readings. This increases the uncertainty of sensor data, as such the reliability of recognition.

Thirdly, most ADLs are composed of a sequence of temporally related actions. As such, sensor data related to an ADL is generated incrementally as the ADL unfolds. In order to provide context-aware assistance for an SH inhabitant, activity recognition should be performed at discrete time points in real-time in a progressive manner. This will accommodate the ever-changing sensor data to recognise the current state of the ongoing activity and further identify the user's needs at the correct time.

1.6 Research Trends and Directions

1.6.1 Complex Activity Recognition

Most existing work on activity recognition is built upon simplified use scenarios, normally focusing on single-user single-activity recognition. In real world situations, human activities are often performed in complex manners. These include, for example, that a single actor performs interleaved and concurrent activities, and/or a group of actors interact with each other to perform joint activities. The approaches and algorithms described in previous sections cannot be applied directly to these application scenarios. Researchers in related communities have realised this knowledge gap and more attention is being focused towards this area as depicted in Fig. 1.3. This shift of research emphasis is also driven by the increasing demand on scalable solutions that are deployable to real world use cases. Nevertheless, research endeavours in this niche field are still at infancy.

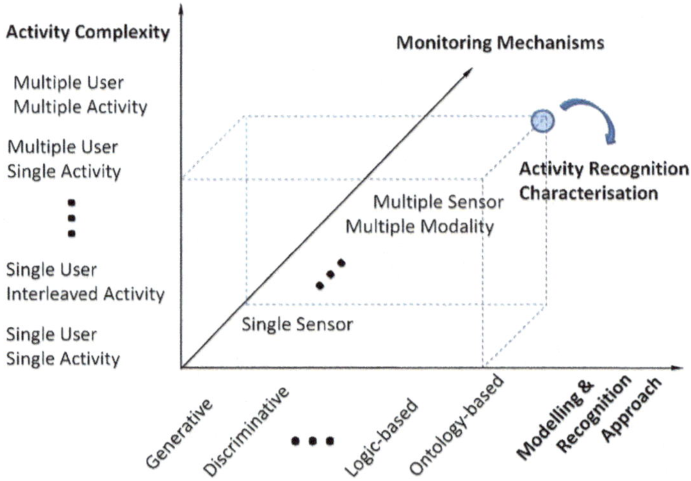

Fig. 1.3 A three-dimensional characterization for activity recognition

In the modelling and recognition of complex activities of a single user, Wu et al. [30] proposed an algorithm using the factorial conditional random field (FCRF) for recognising multiple concurrent activities. This model can handle concurrency but cannot model interleaving activities and cannot be easily scaled up. Hu et al. [30] proposed a two-level probabilistic and goal-correlation framework that deals with both concurrent and interleaving goals from observed activity sequences. They exploited skip-chain conditional random fields (SCCRF) at the lower level to estimate the probabilities of whether each goal is being pursued given a newly observed activity. At the upper level, they used a learnt graph model of goals to infer goals in a "collective classification" manner. Modayil et al. [31] introduced Interleaved Hidden Markov Models to model both inter-activity and intra-activity dynamics. To reduce the size of the state space, they used an approximation for recognising multitasked activities. Gu et al. [32] proposed an Emerging Patterns based approach to Sequential, Interleaved and Concurrent Activity Recognition (epSICAR). They exploit Emerging Patterns as powerful discriminators to differentiate activities. Different from other learning-based models built upon the training dataset for complex activities, they built activity models by mining a set of Emerging Patterns from the sequential activity trace only and applied these models in recognising sequential, interleaved and concurrent activities.

In the modelling and recognition of complex activities of a group or multiple occupants, existing work has mainly focused on vision analysis techniques for activity recognition from video data. Various HMM models have been developed for modelling an individual person's behaviour, interactions and probabilistic data associations. These include the dynamically multi-linked HMM model [33], the hierarchical HMM model [34], the Coupled HMM [35], the mixed-memory Markov model [36] and the Layered Hidden Markov Models (LHMMs) [37]. DBN models are also extensively used to model human interaction activities [38, 39], both using video cameras. Lian et al. [40] used FCRF to conduct inference and learning from patterns of multiple concurrent chatting activities based on audio streams. Work on using dense sensing for complex activity recognition is rare. Lin et al. [41] proposed a layered model to learn multiple users' activity preferences based on sensor readings deployed in a home environment. Nevertheless, their focus is on learning of preference models of multiple users rather than on recognising their activities. Wang et al. [42] used CHMMs to recognise multi-user activities from dense sensor readings in a smart home environment. They developed a multimodal sensing platform and presented a theoretical framework to recognise both single-user and multi-user activities. Cook et al. [43] proposed a single HMM model for two residents. The model can not only represent transitions between activities performed by one person but also represent transitions between residents and transitions between different activities performed by different residents. As such their probabilistic models of activities are able to recognise activities in complex situations where multiple residents are performing activities in parallel in the same environment. Hoey and Grzes [44] present a multi-level DBN model for providing assistance in multiple simultaneous tasks, where the recognition of activities is implicitly used to tailor assistance based on the recognised activities of the person.

While increasing attention has been drawn into this area, nevertheless, research in this niche field is still at its infancy. With the intensive interest in related areas such as smart environments, pervasive computing and novel applications, it is expected that research on activity recognition along this dimension will continue to receive attention and generate results in the next few years.

1.6.2 Domain Knowledge Exploitation

As can be seen from the above discussions, at present, there is a multitude of sensing technologies, multimodal devices and communication platforms being developed and deployed in smart environments for activity monitoring. There is an abundance of approaches and algorithms for activity recognition in various scenarios, including a single user performing a single activity, a single user performing interleaved multiple activities and multiple users performing complex activities. Nevertheless, existing endeavours for activity monitoring and recognition suffer from several main drawbacks.

Firstly, sensor data generated from activity monitoring, in particular in the situations of using multimodal sensors and different types of sensors, are primitive and heterogeneous in format and storage, and separated from each other in both structure and semantics. Such data sets are usually ad hoc, lack of descriptions, thus difficult for exchange, sharing and reuse. To address this problem researchers have made use of domain knowledge to develop high-level formal data models. Nugent et al. [45] proposed a standard XML schema HomeML for smart home data modelling and exchange; Chen et al. [46] proposed context ontologies to provide high-level descriptive sensor data models and related technologies for semantic sensor data management aiming to facilitate semantic data fusion, sharing and intelligent processing. We believe knowledge rich data modelling and standardisation supported by relevant communities is a promising direction towards a commonly accepted framework for sensor data modelling, sharing and reuse.

Secondly, current approaches and algorithms for activity recognition are often carefully handcrafted to well-defined specific scenarios. Existing implemented proof-of-concept systems are mainly accomplished by plumbing and hardwiring the fragmented, disjointed, and often ad hoc technologies. This makes these solutions subject to environment layout, sensor types and installation, and specific application scenarios, i.e., lack of interoperability and scalability. The latest experiments performed by Biswas et al. [47] indicated it is difficult to replicate and duplicate a solution in different environments even for the same, simplest single-user single-activity application scenario. This highlights the challenge to generalise approaches and algorithms of activity recognition to real world use cases. While it is not realistic to pre-define one-size-fits-all activity models due to the number of activities and the variation of the way activities are performed, it is desirable if rich domain knowledge can be exploited to produce initial explicit generic activity models. These models are later used, on the one hand, to generate fine-grained individual-specific activity models, and on the other hand, to evolve towards completion through learning. Chen et al. [48] developed activity ontologies where concepts represent the course-grained activity models while instances represent user activity profiles. Okeyo et al. [49] extended this idea by developing learning algorithms to automatically create fine-grained individual-specific activity models, and also learn new activity models to evolve ontologies towards model completion. Initial results are promising, and further work is needed along this line.

1.6.3 Multi-level Activity Modelling for Scalability and Reusability

Current approaches and algorithms for activity recognition are often carefully hand-crafted to well-defined specific scenarios, both activities and the environment. Existing implemented proof-of-concept systems are mainly accomplished by plumbing and hardwiring the fragmented, disjointed, and often ad hoc technologies. This makes these solutions subject to environment layout, sensor types and installation, and specific activities and users. The solutions thus suffer from a lack of interoperability and scalability. The latest experiments performed by Biswas et al. [47] indicated it is difficult to replicate and duplicate a solution in different environments even for the same, simplest single-user single-activity application scenario. This highlights the challenge to generalize approaches and algorithms of activity recognition to real world use cases.

While it is not realistic to pre-define one-size-fits-all activity models due to the number of activities and the variation of the way activities are performed, we believe multi-level activity modelling and corresponding inference mechanisms at different levels of details could address the problem of applicability and scalability. The basic idea is similar to the concepts of class and object in object-oriented programming (OOP) in that an activity are modelled at both coarse-grained and fine-grained levels. The coarse-grained level activity models are generic like a class in OOP that can be used by any one in many application scenarios. The fine-grained activity models are user specific like an instance in OOP that can accommodate the preference and behaviour habits of a particular user. As such, the models and associated recognition methods can be applied to a wide range of scenarios. Chen et al. [48] developed activity ontologies where concepts represent the course-grained activity models while instances represent user activity profiles. Okeyo et al. [49] extended this idea by developing learning algorithms to automatically create fine-grained individual-specific activity models, and also learn new activity models to evolve ontologies towards model completion. Initial results are promising, and further work is needed along this line. A systematic view on activity recognition.

Research on activity recognition has increasingly taken a systematic view, which takes into consideration seamless integration and reuse of activity modelling, representation, inference and application-specific features. Some of the existing systems have integrated activity modelling with decision making using artificial intelligence techniques [50]. In data driven approaches, the most successful of these are probably the decision theoretic models that extend the DBNs by adding utilities and actions. These models are known as Markov decision processes (MDPs) and their partially observable counterparts POMDPs. These models have been used for scheduling [51] and for assistance with a variety of activities of daily living [52, 53]. POMDPs have also been integrated with high-level knowledge from psychological analyses to generate prompting system for tasks involving dense sensing [54]. In knowledge driven approaches, ontologies have been used as a conceptual backbone for modelling and representation of sensors, context and activities, ranging from context management, data integration, interoperation and sharing, to activity recognition, decision-making support [48, 55–57]. As technologies are adaptive towards real world use cases and transformed into products, it is expected the systematic view on activity recognition will get growing currency.

Domain knowledge will certainly play a dominant role when activity recognition is designed as a component of a complete system, e.g., as an input to support inference and decision making. An envisioned application is to use activity recognition to perform behavioural or functional assessment of adults in their everyday environments. This type of automated assessment also provides a mechanism for evaluating the effectiveness of alternative health interventions. For example, Patel et al. [58] used accelerometer sensor data to analyse and assess the activities of patients with Parkinson's disease. They developed analysis metrics and compared the results with assessment criteria from domain experts to estimate the severity of symptoms and motor complications. This demonstrates that domain knowledge about activity profiling and assessment heuristics are valuable for providing automated health monitoring and assistance in an individual's everyday environment.

1.6.4 Infrastructure Mediated Activity Monitoring

The current practice of installing a large number of sensors and an extensive sensing infrastructure in an environment, e.g., residential homes, has been widely viewed as problematic in real world scenarios. A recent trend for activity monitoring is to leverage a home's existing infrastructure to "reach into the home" with a small set of strategically-placed sensors [59]. The residential infrastructure is usually referred to as existing water pipes, electrical circuits, meters, heating or venting passages or facilities within a home. The basic idea is to capture parameter or state changes of an infrastructure from which activities can be monitored and further recognised. Infrastructure mediated activity monitoring requires selecting appropriate sensors and designing elaborate methods for combining the sensors with the existing infrastructure. Patel et al. [60] detected human movement by differential air pressure sensing in HVAC system ductwork. Fogarty et al. [59] deploy a small number of low-cost sensors at critical locations in a home's existing water distribution infrastructure. The authors infer activities in the home based on water usage patterns. Infrastructure mediated activity monitoring is a very promising direction for activity monitoring and recognition, and certainly worth further exploration.

1.6.5 Intent or Goal Recognition

Current activity recognition and its application is roughly a bottom-up approach starting from the lowest sensor data, then discovering the activity and purposes of the user through increasing higher-level processing. An emerging trend is to adopt a top-down approach to activity monitoring and recognition. namely to (1) recognise or discover the intent or goal of a user, (2) identify the activity that can achieve the goal, (3) monitor the user's behaviour including the performed actions, (4) decide whether or not the user is doing the right thing in terms of the activity model and the monitored behaviour, and finally (5) provide personalized context-aware interactions or assistance whenever and wherever needed. Goals can be either explicitly manually specified, such as when a care provider defines goals for a patient to achieve during a day, or learnt based on domain context. Activities are pre-defined in some flexible way and linked to specific goals. As such, once a goal is specified or identified, applications can instruct/remind users to perform the corresponding activity.

While research on cognitive computation, goal modelling and representation of motivations, goals, intention, belief and emotion, has been undertaken widely in AI communities, in particular within intelligent agent research, the adoption of the knowledge and research results in pervasive computing and smart environments and their applications have so far receive little attention [61]. Nevertheless, interest is growing and a recent workshop on situation, activity and goal awareness (SAGAware) has been organised [62], aiming to facilitate knowledge transfer and synergy, bridge gaps between different research communities/groups, lay down a foundation for common purposes, and help identify opportunities and challenges.

1.6.6 Abnormal Activity Recognition

The existing research on activity recognition focuses mainly on normal activities that may account for the majority of collected data and processing computation. Nevertheless, the results may contribute significantly less towards the purposes of activity recognition as most applications involving activity recognition intend to detect abnormal activities. This is a particularly important task in security monitoring where suspicious activities need to be dealt with and healthcare applications where assistance need to be provided for incapable users. While this view may generate cost-effective results, solving the problem is challenging. Firstly, the concept of an abnormal activity has not been well defined and elaborated with a variety of interpretations available. For instance, everyone performs activity *A*, one person carries out activity *B*. there are different views on whether or not activity *B* is abnormal. Yin et al. [63] defined abnormal activities as events that occur rarely and have not been expected in advance. Secondly, there is an unbalanced data problem in abnormal activity detection. A much larger proportion of sensing data is about normal activity, while the data for abnormal ones are extremely scarce, which makes training the classification model quite difficult. Knowledge driven approaches can certainly fit in. The problem is really about the completeness of priori domain knowledge. For example, is it possible to predict the behaviour of a terrorist in advance or based on previous experience? Clearly, a raft of research problems and issues are open for further investigation.

1.6.7 Sensor Data Reuse and Repurposing

Currently, sensor data generated from activity monitoring, in particular in the situations of using multimodal sensors and different types of sensors, are primitive and heterogeneous in format and storage, and separated from each other in both structure and semantics. Such data sets are usually ad hoc, lack of descriptions, thus difficult for exchange, sharing and reuse. To address these problems researchers have made use of priori domain knowledge to develop high-level formal data models. Nugent et al. [45] proposed a standard XML schema HomeML for smart home data modelling and exchange; Chen et al. [46] proposed context ontologies to provide high-level descriptive sensor data models and related technologies for semantic sensor data management aiming to facilitate semantic data fusion, sharing and intelligent processing. We believe knowledge rich data modelling and standardization supported by relevant communities is a promising direction towards a commonly accepted framework for sensor data modelling, sharing and repurposing. This idea is also in line with the infrastructure mediated monitoring, namely deploy-once, and reuse-for-all.

1.7 Summary

There is no doubt that cyber-physical systems in smart environments will pervade future working and living spaces, transform our lives and impact our society. Activity recognition is becoming an increasingly important determinant to the success of context-aware personalised pervasive applications. Synergistic research efforts in various scientific disciplines, e.g., computer vision, artificial intelligence, sensor networks and wireless communication to name but a few, have brought us a diversity of approaches and methods to address this issue. In this chapter, we first present the background of the research on activity recognition for cyber-physical systems in smart environments. Then we introduce the concept and rationale of activity recognition, and further characterise activity recognition in terms of the way activities are monitored, modelled and analysed. In addition, it highlights a number of application areas for which activity recognition plays key roles, and further describes the typical application scenario of ambient assisted living, as well as challenges of activity recognition in assisted living within a smart home. Followed the above descriptions, we discuss existing research trends and directions in this area. This chapter serves the purpose of establishing the overall picture of activity recognition and further pointing to core constituent components, key research issues, directions and tends. It sets up the scene for readers to proceed to following chapters.

References

1. Weiser M (1991) The computer for the 21st century. Sci Am (1991)
2. Turaga P, Chellappa R, Subrahmanian VS, Udrea O (2008) Machine recognition of human activities: a survey. IEEE Trans Circuits Syst Video Technol
3. Wren CR, Tapia EM (2006) Toward scalable activity recognition for sensor networks. In: Lecture notes in computer science (including subseries Lecture notes in artificial intelligence and lecture notes in bioinformatics)
4. Wang Y, Cang S, Yu H (2018) A data fusion-based hybrid sensory system for older people's daily activity and daily routine recognition. IEEE Sens J 18:6874–6888
5. Aggarwal JK, Cai Q (1999) Human motion analysis: a review. Comput Vis Image Underst
6. Cédras C, Shah M (1995) Motion-based recognition a survey. Image Vis Comput
7. Gavrila DM (1999) The visual analysis of human movement: a survey. Comput Vis Image Underst
8. Poppe R (2010) A survey on vision-based human action recognition. Image Vis Comput
9. Moeslund TB, Hilton A, Krüger V (2006) A survey of advances in vision-based human motion capture and analysis. Comput Vis Image Underst 104(2–3):90–126
10. Yilmaz A, Javed O, Shah M (2006) Object tracking: a survey. ACM Comput Surv
11. Weinland D, Ronfard R, Boyer E (2011) A survey of vision-based methods for action representation, segmentation and recognition. Comput Vis Image Underst
12. Aggarwal JK, Ryoo MS (2011) Human activity analysis: a review. ACM Comput Surv
13. MIT: House_n The PlaceLab. http://web.mit.edu/cron/group/house_n/placelab.html
14. Helal S, Mann W, El-Zabadani H, King J, Kaddoura Y, Jansen E (2005) The gator tech smart house: a programmable pervasive space. Computer (Long Beach Calif) 38:50–60
15. Design I Inhaus design. http://inhausdesign.co.uk/projects/
16. Georgia Institute of Technology: Aware Home Research Initiative. http://www.awarehome.gatech.edu/
17. The Domus Laboratory. https://cswww.essex.ac.uk/iieg/idorm.htm
18. The IDORM project. https://www.usherbrooke.ca/domus/fr/
19. Lin W, Xing S, Nan J, Wenyuan L, Binbin L (2018) Concurrent recognition of cross-scale activities via sensorless sensing. IEEE Sens J 19(2):658–669
20. De-La-Hoz-Franco E, Ariza-Colpas P, Quero JM, Espinilla M (2018) Sensor-based datasets for human activity recognition – a systematic review of literature. IEEE Access 6:59192–59210
21. Nef T, Urwyler P, Büchler M, Tarnanas I, Stucki R, Cazzoli D, Müri R, Mosimann U (2015) Evaluation of three state-of-the-art classifiers for recognition of activities of daily living from smart home ambient data. Sensors (Basel) 15:11725–11740
22. Liu J, Shahroudy A, Xu D, Kot Chichung A, Wang G (2017) Skeleton-based action recognition using spatio-temporal LSTM network with trust gates. IEEE Trans Pattern Anal Mach Intell XX:1–14
23. Liao L, Fox D, Kautz H (2007) Extracting places and activities from GPS traces using hierarchical conditional random fields. Int J Rob Res
24. Huỳnh T, Schiele B (2006) Unsupervised discovery of structure in activity data using multiple eigenspaces. In: Hazas M, Krumm J, Strang T (eds) Location- and context-awareness. Springer, Berlin, pp 151–167

25. Kautz HA (2014) A formal theory of plan recognition and its implementation. In: Reasoning about plans
26. Wobcke W (2002) Two logical theories of plan recognition. J Log Comput
27. Bouchard B, Giroux S, Bouzouane A (2006) A smart home agent for plan recognition of cognitively-impaired patients. J Comput
28. Chen L, Nugent C, Mulvenna M, Finlay D, Hong X, Poland M (2008) A logical framework for behaviour reasoning and assistance in a smart home. Lecture notes in computer science (including subseries Lecture notes in artificial intelligence and lecture notes in bioinformatics)
29. Triboan D, Chen L, Chen F, Wang Z (2017) Semantic segmentation of real-time sensor data stream for complex activity recognition. Pers Ubiquitous Comput
30. Wu T, Lian C, Hsu JYY (2007) Joint recognition of multiple concurrent activities using factorial conditional random fields. In: Proceedings 22nd conferences on artificial intelligence
31. Modayil J, Bai T, Kautz H (2008) Improving the recognition of interleaved activities. In: Proceedings of the 10th international conference on Ubiquitous computing - UbiComp '08
32. Gu T, Wu Z, Tao X, Pung HK, Lu J (2009) epSICAR: an emerging patterns based approach to sequential, interleaved and concurrent activity recognition. In: 7th annual IEEE international conference on pervasive computing and communications, PerCom 2009
33. Gong S, Xiang T (2003) Recognition of group activities using dynamic probabilistic networks. In: Proceedings ninth IEEE international conference on computer vision
34. Nguyen N, Venkatesh S, Bui H (2006) Recognising behaviours of multiple people with hierarchial probabilistic model and statistical data association. In: British machine vision conference
35. Oliver N, Rosario B, Pentland A (1999) A Bayesian computer vision system for modeling human interactions. In: Lecture notes in computer science (including subseries Lecture notes in artificial intelligence and lecture notes in bioinformatics)
36. Choudhury T, Basu S (2005) Modeling conversational dynamics as a mixed-memory markov process. Adv Neural Inf Process Syst 17
37. Oliver N, Garg A, Horvitz E (2004) Layered representations for learning and inferring office activity from multiple sensory channels. In: Computer vision and image understanding
38. Wyatt D, Choudhury T, Bilmes J, Kautz H (2007) A privacy-sensitive approach to modeling multi-person conversations. In: IJCAI international joint conference on artificial intelligence
39. Youtian D, Feng C, Wenli X, Yongbin L (2006) Recognizing interaction activities using dynamic Bayesian network. In: Proceedings - international conference on pattern recognition
40. Lian C-C, Hsu JY-J (2008) Chatting activity recognition in social occasions using factorial conditional random fields with iterative classification. In: Proceedings of the 23rd national conference on artificial intelligence, vol 3. AAAI Press, pp 1814–1815
41. Lin ZH, Fu LC (2007) Multi-user preference model and service provision in a smart home environment. In: Proceedings of the 3rd IEEE international conference on automation science and engineering, IEEE CASE 2007
42. Wang L, Gu T, Tao X, Lu J (2008) Sensor-based human activity recognition in a multi-user scenario. In: Lecture notes in computer science (including subseries Lecture notes in artificial intelligence and lecture notes in bioinformatics)
43. Singla G, Cook DJ, Schmitter-Edgecombe M (2010) Recognizing independent and joint activities among multiple residents in smart environments. J Ambient Intell Humaniz Comput
44. Hoey J, Grze M (2011) Distributed control of situated assistance in large domains with many tasks. In: Twenty-first international conference on automated planning and scheduling (2011)
45. Nugent C, Finlay D, Davies R, Wang H, Zheng H, Hallberg J, Synnes K, Mulvenna M (2007) homeML ? An open standard for the exchange of data within smart environments. In: Pervasive computing for quality of life enhancement
46. Chen L, Nugent C, Al-Bashrawi A (2009) Semantic data management for situation-aware assistance in ambient assisted living. In: Proceedings of the 11th international conference on information integration and web-based applications & services - iiWAS '09
47. Biswas J, Baumgarten M, Tolstikov A, Wai AAP, Nugent C, Chen L, Donnelly M (2010) Requirements for the deployment of sensor based recognition systems for ambient assistive living. In: Lecture notes in computer science (including subseries Lecture notes in artificial intelligence and lecture notes in bioinformatics)

48. Chen L, Nugent C (2009) Ontology-based activity recognition in intelligent pervasive environments. Int J Web Inf Syst
49. Okeyo G, Chen L, Wang H, Sterritt R (2011) Ontology-based learning framework for activity assistance in an adaptive smart home. In: Chen L, Nugent CD, Biswas J, Hoey J (eds) Activity recognition in pervasive intelligent environments. Atlantis Press, Paris, pp 237–263
50. Modayil J, Levinson R, Harman C (2008) Integrating sensing and cueing for more effective activity reminders. In: AAAI fall symposium: AI in eldercare: new solutions to old problems
51. Pollack ME, Brown L, Colbry D, McCarthy CE, Orosz C, Peintner B, Ramakrishnan S, Tsamardinos I (2003) Autominder: an intelligent cognitive orthotic system for people with memory impairment. In: Robotics and autonomous systems
52. Kan P, Huq R, Hoey J, Goetschalckx R, Mihailidis A (2011) The development of an adaptive upper-limb stroke rehabilitation robotic system. J Neuroeng Rehabil 8:33
53. Hoey J, Poupart P, von Bertoldi A, Craig T, Boutilier C, Mihailidis A (2010) Automated handwashing assistance for persons with dementia using video and a partially observable Markov decision process. Comput Vis Image Underst
54. Hoey J, Pltz T, Jackson D, Monk A, Pham C, Olivier P (2011) Rapid specification and automated generation of prompting systems to assist people with dementia. Pervasive Mob Comput
55. Latfi F, Lefebvre B, Descheneaux C (2007) Ontology-based management of the telehealth smart home, dedicated to elderly in loss of cognitive autonomy. In: CEUR workshop proceedings (2007)
56. Klein M, Schmidt A, Lauer R (2007) Ontology-centred design of an ambient middleware for assisted living: the case of SOPRANO. Context (2007)
57. Chen L, Nugent C, Mulvenna M, Finlay D, Hong X (2009) Semantic smart homes: towards knowledge rich assisted living environments. Stud Comput Intell
58. Patel S, Lorincz K, Hughes R, Huggins N, Growdon J, Standaert D, Akay M, Dy J, Welsh M, Bonato P (2009) Monitoring motor fluctuations in patients with parkinsons disease using wearable sensors. IEEE Trans Inf Technol Biomed (2009)
59. Fogarty J, Au C, Hudson SE (2006) Sensing from the basement: a feasibility study of unobtrusive and low-cost home activity recognition. In: Proceedings of the annual ACM symposium on user interface software and technology
60. Patel SN, Reynolds MS, Abowd GD (2008) Detecting human movement by differential air pressure sensing in HVAC system ductwork: an exploration in infrastructure mediated sensing. In: Lecture notes in computer science (including subseries Lecture notes in artificial intelligence and lecture notes in bioinformatics)
61. Mei L, Easterbrook S (2007) Evaluating user-centric adaptation with goal models. In: Proceedings - ICSE 2007 workshops: first international workshop on software engineering for pervasive computing applications, systems, and environments, SEPCASE'07
62. Chen L, Rashidi P (2012) Situation, activity and goal awareness in ubiquitous computing. Int J Pervasive Comput Commun 8:216–224
63. Yin J, Yang Q, Pan JJ (2008) Sensor-based abnormal human-activity detection. IEEE Trans Knowl Data Eng

Chapter 2
Sensor-Based Activity Recognition Review

2.1 Introduction

The idea of using sensors for activity monitoring and recognition has been existent since the late 90s. It was initially pioneered and experimented by the work of the Neural Network house [1] in the context of home automation, and a number of location-based applications aiming to adapt systems to users' whereabouts [2, 3]. The approach was soon found to be more useful and suitable in the area of ubiquitous and mobile computing—an emerging area in the late 90s, due to its easy deployment. As such, extensive research has been undertaken to investigate the use of sensors in various application scenarios of ubiquitous and mobile computing, leading to considerable work on context-awareness [4–6], smart appliances [40, 41] and activity recognition [7–10]. Most research at that time made use of wearable sensors, either dedicated sensors attached to human bodies or portable devices like mobile phones, with application to ubiquitous computing scenarios such as providing context-aware mobile devices. Activities being monitored in these researches are mainly physical activities like motion, walking and running. These early works lay a solid foundation for wearable computing and still inspire and influence today's research.

In the early 2000s, a new sensor-based approach that uses sensors attached to objects to monitor human activities appeared. This approach, which was later dubbed as the "dense sensing" approach, performs activity recognition through the inference of user-object interactions [11, 12]. The approach is particular suitable for dealing with activities that involve a number of objects within an environment, or instrumental Activities of Daily Living [13, 14]. Research on this approach has been heavily driven by the intensive research interests and huge research effort on smart home-based assistive living, such as the EU's AAL program [15]. In particular, sensor-based activity recognition can better address sensitive issues in assistive living such as privacy, ethics and obtrusiveness than conventional vision-based approaches. This combination of application needs and technological advantages has stimulated considerable research activities in a global scale, which gave rise to a large number of research projects, including the House_n, CASAS, Gator-Tech, inHaus, AwareHome,

L. Chen and C. D. Nugent, *Human Activity Recognition and Behaviour Analysis*, https://doi.org/10.1007/978-3-030-19408-6_2

DOMUS and iDorm projects, to name but a few. As a result of the wave of intensive investigation, there have seen a plethora of impressive works on sensor-based activity recognition in the past several years.

While substantial research has been undertaken, and significant progress has been made, the two main approaches, wearable sensors based and dense sensing based activity recognition are currently still focuses of study. The former is mainly driven by the ever-popular pervasive and mobile computing while the latter is predominantly driven by smart environment applications such as ambient assisted living. Interests in various novel applications are still increasing and application domains are rapidly expanding.

2.2 Sensor-Based Activity Monitoring

Currently a wide range of sensors, including contact sensors, RFID, accelerometers, audio and motion detectors, to name but a few, are available for activity monitoring. These sensors are different in types, purposes, output signals, underpinning theoretical principles and technical infrastructure. However, they can be classified into two main categories in terms of the way they are deployed in activity monitoring applications. These are wearable sensors and dense sensors, and are described in details in the following.

2.2.1 Wearable Sensor Based Activity Monitoring

Wearable sensors generally refer to sensors that are positioned directly or indirectly on a human body. They generate signals when the user performs activities. As a result, they can monitor features that are descriptive of the person's physiological state or movement. Wearable sensors can be embedded into clothes, eyeglasses, belts, shoes, wristwatches, mobile devices or positioned directly on the body. They can be used to collect information such as body position and movement, pulse, and skin temperature. Researchers have found that different types of sensor information are effective for classifying different types of activities. In the following, we summarise the common practice in wearable sensor-based activity monitoring.

Accelerometer sensors are probably the most frequently used wearable sensor for activity monitoring. They are particularly effective in monitoring actions that involve repetitive body motions, such as walking, running, sitting, standing, climbing stairs. Bao et al. [11] provide a summary of research work that recognises human activities using acceleration data. Kern et al. [16] deploy a network of 3-axis accelerometers distributed over the user's body. Each accelerometer provides information about the orientation and movement of the corresponding body part. Lukowicz et al. [17] recognize workshop activities using body-worn microphones and accelerometers. Measuring acceleration and angular velocity (the angle of the user's thigh) through wearable sensors, such as accelerometers and gyroscopes, Lee et al. [10] propose a dead-reckoning method for determining a user's location and recognizing sitting, standing and walking behaviours. Mantyjarvi [18] recognises human ambulation and posture on acceleration data collected from the hip.

GPS sensors are another widely used wearable sensor for monitoring location-based activities in open pervasive and mobile environments. Patterson et al. [12] present details of detecting human high-level behaviour from a GPS sensor stream, such as boarding a bus at a particular bus stop, travelling and disembarking. Ashbrook et al. [19] use GPS to learn significant locations and predict movement across multiple users. Liao et al. [20] learn and infer a user's mode of transportation and their goal in addition to abnormal behaviours (e.g., taking a wrong bus) based on GPS data logs.

Biosensors are an emerging technology aiming to monitor activities through vital signs. A diversity of sensors in different forms has been studied in order to measure the wide range of vital signs such as blood pressure, heart rate, EEG, ECG and respiratory information. Sung et al. [21] monitor the body temperature of soldiers to detect hypothermia. Harms et al. [22] use information gathered by a smart garment to identify body posture.

In addition to the investigation of different wearable sensors for activity monitoring, research on the support and novel application of wearable computing has been undertaken. Pantelopoulos et al. [23] present a survey on wearable systems for monitoring and early diagnosis for the elderly. Dakopoulos and Bourbakis [24] present a survey on wearable obstacle avoidance electronic travel aids for visually impaired. Yoo et al. [25] design on-body and near-body networks that use the human body itself as a channel for creating BodyNets. Cooper and Au use wearable sensors to design and evaluate assistive wheelchairs [26] and smart walking sticks [27]. Kim et al. [28] use wearable sensors to recognize gestures. Madan et al. [29] characterize a person's social context by evaluating a user's proximity, speech, head movements and galvanic skin response.

Wearable sensor-based activity monitoring suffers from limitations. Most wearable sensors need to run continuously and be operated hands-free. This may have difficulties in real-world application scenarios. Practical issues include the acceptability or willingness to use wearable sensors and the viability and ability to wear them. Technical issues include the size, ease of use, battery life and effectiveness of the approach in real-world scenarios. To address these issues, vigorous investigation of smart garments has been carried out, which aims to embed sensors in garments for monitoring [30]. Another research thread is to make use of existing gadgets that have already been carried in a daily basis like smartphones as intelligent sensors for activity monitoring, recognition and assistance. This practice has been in place for a while and is expected to gain large-scale uptake given the latest development and affordability of such palm-held electronic devices.

Obviously, wearable sensors are not suitable for monitoring activities that involve complex physical motions and/or multiple interactions with the environment. In some cases, sensor observations from wearable sensors alone are not sufficient to differentiate activities involving simple physical movements (e.g., making tea and making coffee). As a result, dense sensing based activity monitoring has emerged, which is described below.

2.2.2 *Ambient Sensor Based Activity Monitoring*

Ambient sensor based activity monitoring refers to the practice that a large number of ambient sensors are attached to objects within an environment, and activities are monitored by detecting user-object interactions. The approach is based on real-world observations that activities are characterized by the objects that are manipulated during their performance. A simple indication of an object being used can often provide powerful clues about the activity being undertaken. As such it is assumed that activities can be recognised from sensor data that monitors human interactions with objects in the environment. By dense sensing, we refer to the way and scale with which sensors are used. Using dense sensing a large number of sensors, normally low-cost low-power and miniaturized, are deployed in a range of objects or locations within an environment for the purpose of monitoring movement and behaviour.

As dense sensing-based monitoring embeds sensors within environments, this makes it more suitable for creating ambient intelligent applications such as smart environments. As such, dense sensing-based activity monitoring has been widely adopted in ambient assisted living (AAL), via the smart home paradigm [14]. Sensors in an SH can monitor an inhabitant's movements and environmental events so that assistive agents can infer the undergoing activities based on the sensor observations, thus providing just-in-time context-aware ADL assistance. For instance, a switch sensor in the bed can strongly suggest sleeping, and pressure mat sensors can be used for tracking the movement and position of people within the environment.

Since the introduction of the idea in the early 2000s, extensive research has been undertaken to investigate the applicability of the approach in terms of sensor types, modalities and applications. For example, Tapia et al. [64] use environmental state-change sensors to collect information about interaction with objects and recognize activities that are of interest to medical professionals such as toileting, bathing, and grooming. Wilson et al. [31] use four kinds of anonymous and binary sensors, motion detectors, break-beam sensors, pressure mats, and contact switches for simultaneous tracking and activity recognition. Wren et al. [32] employ networks of passive infrared motion sensors to detect the presence and movement of heat sources. With this captured data they can recognize low-level activities such as walking, loitering, and turning, as well as mid-level activities such as visiting and meeting. Srivastava et al. [33] exploit wireless sensor network to develop a smart learning environment for young children. Hollosi et al. [34] use voice detection techniques to perform acoustic event classification for monitoring in Smart Homes. Simple object sensors are adopted in [35].

Given the abundance of different types and modalities of sensors, sensors have been used in different ways and combinations for dense sensing activity monitoring in many application scenarios. It is impossible to claim that one sensor deployment for a specific application scenario is superior to the other. The suitability and performance is usually down to the nature of the type of activities being assessed and the characteristics of the concrete applications. As such, in this chapter, we shall not discuss in detail the different usage of dense sensing in various scenarios but simply introduce its rationale as described above.

Generally speaking, wearable sensor-based activity monitoring receives more attention in mobile computing while dense sensing is more suitable for intelligent environment enabled applications. It is worth pointing out that wearable sensors and dense sensing are not mutually exclusive. In some applications, they have to work together. For example, RFID (Radio Frequency Identification) based activity monitoring requires that objects are instrumented with tags and users wear an RFID reader affixed to a glove or a bracelet. Philipose and Fishkin [36, 37] developed two devices, iGlove and iBracelet, working as wearable RFID readers that detect when users interact with unobtrusively tagged objects. Patterson et al. [38] performed fine-grained activity recognition (i.e., not just recognising that a person is cooking but determining what they are cooking) by aggregating abstract object usage. Hodges et al. [39] proposed to identify individuals from their behaviour based on their interaction with the objects they use in performing daily activities. Buettner et al. [40] recognize indoor daily activities by using an RFID sensor network. In most cases, wearable sensors and dense sensing are complementary and can be used in combination in order to yield optimal recognition results. For example, Gu et al. [41] combine wearable sensors and object sensors for collecting multimodal sensor information. Through a pattern-based method, they recognize sequential, interleaved and concurrent activities.

While substantial research has been undertaken, and significant progress has been made, the two main approaches, wearable sensors based and dense sensing-based activity recognition are currently still focuses of study. The former is mainly driven by the ever-popular pervasive and mobile computing while the latter is predominantly driven by smart environment applications such as ambient assisted living. Interests in various novel applications are still increasing and application domains are rapidly expanding.

2.3 Data-Driven Approaches to Activity Modelling and Recognition

Data-driven activity modeling can be classified into two main categories: generative and discriminative. In the generative approach, one attempts to build a complete description of the input or data space, usually with a probabilistic model such as a Bayesian network. In the discriminative approach, one only models the mapping from inputs (data) to outputs (activity labels). Discriminative approaches include many heuristic (rule-based) approaches, neural networks, conditional random fields and linear or non-linear discriminative learning (e.g. support vector machines). In the following, we cover major results using each of these methods.

2.3.1 Generative Methods

The simplest possible generative approach is the naïve Bayes classifier, which has been used with promising results for activity recognition. Naïve Bayes classifiers model all observations (e.g. sensor readings) as arising from a common causal source: the activity, as given by a discrete label. The dependence of observations on activity labels is modelled as a probabilistic function that can be used to identify the most likely activity given a set of observations. Despite the fact that these classifiers assume conditional independence of the features, the classifiers yield good accuracy when large amounts of sample data are provided. Nevertheless, naïve Bayes classifiers do not explicitly model any temporal information, usually considered important in activity recognition.

The Hidden Markov Model (HMM) is probably the most popular generative approach that includes temporal information. A HMM is a probabilistic model with a particular structure that makes it easy to learn from data, to interpret the data once a model is learned, and is both easy and efficient to implement. It consists of a set of hidden (latent) states coupled in a stochastic Markov chain, such that the distribution over states at some time depends only on the values of states at a finite number of preceding times. The hidden states then probabilistically generate observations through a stochastic process. HMMs made their impact initially through use in the speech recognition literature, where latent states correspond to phoneme labels, and observations are features extracted from audio data. HMMs have more recently been adopted as a model of choice in computer vision for modelling sequential (video) data. HMM use a Markov chain over a discrete set of states. A closely relative of the HMM uses continuous states, a model usually referred to as a linear dynamical system (LDS). State estimation in LDSs is better known as a Kalman filter. LDSs have been used with inputs from a variety of sensors for physiological condition monitoring [42] in which a method is also introduced to deal with unmodelled variations in data, one of the major shortcomings of the generative approach.

HMMs form the basis of statistical temporal models. They are, in fact, a special case of the more general dynamic Bayesian networks (DBNs), which are Bayesian networks in which a discrete time index is explicitly represented. Inference and learning in DBNs is simply an application of network propagation in Bayesian networks. DBNs usually make a Markovian assumption, but explicitly represent conditional independencies in the variables, allowing for more efficient and accurate inference and learning. A well-known early use of DBNs for activity monitoring was in the Lumière project, where a Microsoft Windows user's need for assistance was modelled based on their activities on the screen [43].

A simple DBN extension of HMMs is the coupled HMM for recognition of simultaneous human actions. Coupled Hidden Markov Models (CHMMs) have two Markovian chains, each modelling a different stream of data, with a coupling between them to model their inter-dependence. Oliver et al. [57] learn a multi-layer model of office activity to choose actions for a computational agent. The model uses multimodal inputs, making only very slight use of computer vision. The Assisted Cognition project [44] has made use of DBNs, in particular for Opportunity Knocks [20], a system designed to provide directional guidance to a user navigating through a city. This system uses a three level hierarchical Markov model represented as a DBN to infer a user's activities from GPS sensor readings. Movement patterns, based on the GPS localization signals, are translated into a probabilistic model using unsupervised learning. From the model and the user's current location, future destinations and the associated mode of transportation can be predicted. Based on the prediction, the system has the ability to prompt the user if an error in route is detected.

Wilson and Atkeson [31] use DBNs to simultaneously track persons and model their activities from a variety of simple sensors (motion detectors, pressure sensors, switches, etc.). DBNs were also used in the iSTRETCH system [45], a haptic robotic device to assist a person with stroke rehabilitation. The DBN models the person's current behaviours, their current abilities, and some aspects of their emotional state (e.g. their responsiveness, learning rate and fatigue level). The person's behaviours correspond to how long they take for each exercise, what type of control they exhibit and whether they compensate. These behaviours are inferred from sensors on the device and in the person's chair.

Even though they are simple and popular, HMMs and DBNs have some limitations. A HMM is incapable of capturing long- range or transitive dependencies of the observations due to its very strict independence assumptions (on the observations). Furthermore, without significant training, a HMM may not be able to recognize all of the possible observation sequences that can be consistent with a particular activity.

2.3.2 Discriminative Methods

A drawback of the generative approach is that enough data must be available to learn the complete probabilistic representations that are required. In this section, we discuss an alternative approach for modelling in which we focus directly on solving the classification problem, rather than on the representation problem. The complete data description of a generative model induces a classification boundary, which can be seen by considering every possible observation and applying the classification rule using inference. The boundary is thus implicit in a generative model, but a lot of work is necessary to describe all the data to obtain it. A discriminative approach, on the other hand, considers this boundary to be the primary objective.

Perhaps the simplest discriminative approach is Nearest Neighbor (NN), in which a novel sequence of observations is compared to a set of template sequences in a training set, and the most closely matching sequences in the training set vote for their activity labels. This simple approach can often provide very good results. Bao and Intille [11] investigated this method along with numerous other base-level classifiers for the recognition of activities from accelerometer data. They found that the simple nearest neighbor approach is outperformed by decision trees, a related method, where the training data is partitioned into subsets according to activity labels and a set of rules based on features of the training data. The rules can then be used to identify the partition (and hence the activity label) corresponding to a new data sample. Maurer et al. [46], employed decision trees to learn logical descriptions of activities from complex sensor readings from a wearable device (the eWatch). The decision tree approach offers the advantage of generating rules that are understandable by the user, but it is often brittle when high precision numeric data is collected. Stikic and Schiele [47] use a clustering method in which activities are considered as a "bag of features" to learn template models of activities from data with only sparse labels.

Many discriminative approaches explicitly take into account the fact that, for classification, it is actually only the points closest to the boundary that are of interest. The ones very far away (the "easy" ones to classify) do not play such a significant role. The challenge is therefore to find these "hard" data points (the ones closest to the boundary). These data points will be known as the "support vectors", and actually define the boundary. A support vector machine (SVM) is a machine learning technique to find these support vectors automatically. A recent example of an SVM in use for activity modelling is presented by Brdiczka et al. [48] where a model of situations is learned automatically from data by first learning roles of various entities using SVMs and labelled training data, then using unsupervised clustering to build 'situations' or relations between entities, which are then labelled and further refined by end users. The key idea in this work is to use a cognitive model (situation model) based on cognitive theory motivated by models of human perception of behaviour in an environment. The CareMedia project [49] also uses an SVM to locate and recognize social interactions in a care facility from multiple sensors, including video and audio. The fusion of video and audio allowed 90% recall and 20% precision in identifying interactions including shaking hands, touching, pushing and kicking. The CareMedia project's goals are to monitor and report behaviour assessments in a care home to caregivers and medical professionals.

Ravi et al. also found that SVMs performed consistently well, but also investigated meta-level classifiers that combined the results of multiple base-level classifiers [50]. Features extracted from worn accelerometers are extracted and classified using five different base-level classifiers (decision tables, decision trees, k-nearest neighbors, SVM and Naïve Bayes). The meta-level classifiers are generated through a variety of techniques such as boosting, bagging, voting, cascading and stacking. For recognizing a set of eight activities including standing, walking, running, going up/down stairs, vacuuming and teeth brushing, they found that a simple voting scheme performed the best for three easier experimental settings, whereas boosted SVM performed best for the most difficult setting (test/training separation across users and days).

In practice, many activities may have non-deterministic natures, where some steps of the activities may be performed in any order, and so are concurrent or interwoven. A conditional random field (CRF) is a more flexible alternative to the HMM that addresses such practical requirements. It is a discriminative and generative probabilistic model that represents the dependence of a hidden variable y on an observed variable x. Both HMMs and CRFs are used to find a sequence of hidden states based on observation sequences. Nevertheless, instead of finding a joint probability distribution $p(x,y)$ as the HMM does, a CRF attempts to find only the conditional probability $p(y|x)$. A CRF allows for arbitrary, non-independent relationships among the observation sequences, hence the added flexibility. Another major difference is the relaxation of the independence assumptions, in which the hidden state probabilities may depend on the past and even future observations. A CRF is modelled as an undirected acyclic graph, flexibly capturing any relation between an observation variable and a hidden state. CRFs are applied to the problem of activity recognition in [51] where they are compared to HMMs, but only in a simple simulated domain. Liao et al. [52] use hierarchical CRFs for modelling activities based on GPS data. Hu and Yang [53] use skip-chain CRFs, an extension in which multiple chains interact in a manner reminiscent of the CHMM, to model concurrent and interleaving goals, a challenging problem for activity recognition. Mahdaviani and Choudhury [54] show how semi-supervised CRFs can be used to learn activity models from wearable sensor data.

2.3.3 Heuristic and Other Methods

Many approaches do not fall clearly into discriminative or generative categories, but rather use a combination of both, along with some heuristic information. The Independent Lifestyle Assistant (ILSA) is an example, as it uses a combination of heuristic rules and statistical models of sequential patterns of sensor firings and time intervals to help a person with planning and scheduling [55]. PEAT (the Planning and Execution Assistant and Trainer) is a cognitive assistant that runs on a mobile device and helps compensate for executive functional impairment. PEAT uses reactive planning to adjust a user's schedule based on their current activities. Activity recognition in PEAT is based on what the user is doing, and on data from sensors on the mobile device. These are fed into an HMM, the outputs of which are combined with the reactive planning engine [56].

Other work has investigated how activities can be modelled with a combination of discriminative and generative approaches [57], how common sense models of everyday activities can be built automatically using data mining techniques [58], and how human activities can be analysed through the recognition of object use, rather than the recognition of human behaviour [59]. This latter work uses DBNs to model various activities around the home, and a variety of radio frequency identification (RFID) tags to bootstrap the learning process. Some authors have attempted to compare discriminative and generative models [11, 50], generally finding the discriminative models yield lower error rates on unseen data, but are less interpretable. Gu et al. [41] use the notion of emerging patterns to look for frequent sensor sequences that can be associated with each activity as an aid for recognition. Omar et al. [60] present a comparative study of a variety of classification methods for analysing multi-modal sensor data from a smart walker.

The generative approach, which attempts to build a complete description of the input or data space, usually with probabilistic analysis methods such as Markov models [61] and Bayesian networks [48] for activity modelling. These methods incorporate an inhabitant's preferences by tuning the initial values of the parameters of the probabilistic models. The major disadvantage with such methods is that the model is static and subjective in terms of probabilistic variable configuration. An alternative approach is referred to as the discriminative approach, which only models the mapping from inputs (data) to outputs (activity labels). Discriminative approaches include many heuristics (rule-based) approaches, for example, neural networks, linear or non-linear discriminant learning. They use machine learning techniques to extract ADL patterns from observed daily activities, and later use the patterns as predictive models [48]. Both approaches require large datasets for training models, thus suffer from the data scarcity or the "Cold Start" problem. It is also difficult to apply modelling and learning results from one person to another.

2.4 Knowledge-Driven Approaches to Activity Modelling and Recognition

Knowledge-driven activity recognition and modelling is motivated by real-world observations that for most activities of daily living and working, the list of objects required for a particular activity is limited and functionally similar. Even if the activity can be performed in different ways the number and type of these involved objects do not vary significantly. For example, it is common sense that the activity "*make coffee*" consists of a sequence of actions involving a *coffee pot, hot water, a cup, coffee, sugar* and *milk*; the activity "*brush teeth*" contains actions involving a *toothbrush, toothpaste, water tap, cup* and *towel*. On the other hand, as humans have different lifestyles, habits or abilities, they may perform various activities in different ways. For instance, one may like strong *white coffee,* and another may prefer a special brand of coffee. Even for the same type of activity (e.g., making *white coffee*), different individuals may use different items (e.g., *skimmed milk* or *whole milk*) and in different orders (e.g., adding *milk* first and then *sugar*, or vice versa). Such domain-dependent activity-specific prior knowledge provides valuable insights into how activities can be constructed in general and how they can be performed by individuals in specific situations.

Similarly, knowledge-driven activity recognition is founded upon the observations that most activities, in particular, routine activities of daily living and working, take place in a relatively specific circumstance of time, location and space. The space is usually populated with events and entities pertaining to the activities, forming a specific environment for specific purposes. For example, brushing teeth is normally undertaken twice a day in a bathroom in the morning and before going to bed and involves the use of toothpaste and a toothbrush; meals are made in a kitchen with a cooker roughly three times a day. The implicit relationships between activities, related temporal and spatial context and the entities involved (objects and people) provide a diversity of hints and heuristics for inferring activities.

Knowledge-driven activity modelling and recognition intends to make use of rich domain knowledge and heuristics for activity modelling and pattern recognition. The rationale is to use various methods, in particular, knowledge engineering methodologies and techniques, to acquire domain knowledge. The captured knowledge can then be encoded in various reusable knowledge structures, including activity models for holding heuristics and prior knowledge in performing activities, and context models for holding relationships between activities, objects and temporal and spatial contexts. Comparing to data-driven activity modelling that learns models from large-scale datasets and recognises activities through data intensive processing methods, knowledge-driven activity modelling avoids a number of problems, including the requirement for large amounts of observation data, the inflexibility that arises when each activity model needs to be computationally learned, and the lack of reusability that results when one person's activity model is different from another's.

Knowledge structures can be modelled and represented in different forms, such as schemas, rules or networks. This will decide the way and the extent to which knowledge is used for following processing such as activity recognition, prediction and assistance. In terms of the manner in which domain knowledge is captured, represented and used, knowledge-driven approaches to activity modelling and recognition can be roughly classified into three main categories as presented in the following sections.

2.4.1 Mining-Based Approach

The rationale of a mining-based approach is to create activity models by mining existing activity knowledge from publicly available sources. More specifically, given a set of activities, the approach seeks to discover from the text corpuses a set of objects used for each activity and extract object usage information to derive their associated usage probabilities. The approach essentially views the activity model as a probabilistic translation between activity names (e.g., "*make coffee*") and the names of involved objects (e.g., "*mug*", "*milk*"). As the correlations between activities and their objects are common-sense prior knowledge (e.g., most of us know how to carry out daily activities), such domain knowledge can be gleaned in various sources such as how-tos (e.g., those at ehow.com), recipes (e.g., from epicurious.com), training manuals, experimental protocols, and facility/device user manuals.

A mining-based approach consists of a sequence of distinct tasks. Firstly, it needs to identify activities of concern and relevant sources that describe these activities. Secondly, it uses various methods, predominantly information retrieval and analysis techniques, to retrieve activity definitions from specific sources and extract phrases that describe the objects used during the performance of the activity. Then algorithms, predominantly probabilistic and statistical analysis methods such as co-occurrences and association are used to estimate the object-usage probabilities. Finally, the mined object and usage information is used to create activity models such as a HMM that can be used further for activity recognition.

Mining-based activity modelling was initially investigated by researchers from Intel Research [62, 63]. Perkowitz et al. [63] proposed the idea of mining the Web for large-scale activity modelling. They used the QTag tagger to tag each word in a sentence with its part of speech (POS) and a customized regular expression extractor to extract objects used in an activity. They then used the Google Conditional Probabilities (GCP) APIs to determine automatically the probability values of object usage. The mined object and their usage information are then used to construct DBN models through Sequential Monte Carlo (SMC) approximation. They mined the website ehow.com for roughly 2300 directions on performing domestic tasks (from "*boiling water in the microwave*" to "*change your air filter*"), and the website ffts.com and epicurious.com for a further 400 and 18,600 recipes respectively, generating a total 21,300 activity models. Using the DBN activity models they have performed activity recognition for a combination of real user data and synthetic data. While initial evaluation results were positive, the drawback was that there are no mechanisms to guarantee the mined models capturing completely the sequence probabilities and the idiosyncrasy of certain activities. The inability to capture such intrinsic characteristics may limit the model's accuracy in real deployments.

Wyatt et al. [62] followed Perkowitz's approach by mining the Web to create DBN activity models. However, this group extended the work in three aspects, aiming to address the idiosyncrasies and to improve model accuracy. To cover the wide variety of activity definition sources, they mined the Web in a more discriminative way in a wider scope. They did this by building a specialized genre classifier trained and tested with a large number of labelled Web pages. To enhance model applicability, they used the mined models as base activity models and then exploited the Viterbi Algorithm and Maximum Likelihood to learn customized activity parameters from unsegmented, unlabelled sensor data. In a bid to improve activity recognition accuracy they also presented a bootstrap method that produced labelled segmentations automatically. Then they used the Kullback–Leibler (KL) divergence to compute activity similarity.

A difficulty in connecting mined activities with tagged objects is that the activity models may refer to objects synonymously. For example, both a "*mug*" and "*cup*" can be used for making *tea*; both a "*skillet*" and "*frying pan*" be used for making pasta. This leads to a situation that one activity may have different models with each having the same activity name but different object terms. To address this, Tapia et al. [64] proposed to extract collections of synonymous words for the functionally-similar objects automatically from WordNet, an online lexical reference system for the English language. The set of terms for similar objects is structured and represented in a hierarchical form known as the object ontology. With the similarity measure provided by the ontology, an activity model will not only cover a fixed number of object terms but also any other object terms that are in the same class in the ontology.

Another shortcoming of early work in the area is that the segmentation is carried out in sequential order based on the duration of an activity. As the duration of performing a specific activity may vary substantially from one to another, this may give rise to applicability issues. In addition, in sequential segmentation, one error in one segment may affect the segmentations of the subsequent traces. To tackle this, Palmes et al. [65] proposed an alternate method for activity segmentation and recognition. Instead of relying on the order of object use, they exploited the discriminative trait of the usage frequency of objects in different activities. They constructed activity models by mining the Web and extracting relevant objects based on their weights. The weights are then utilized to recognize and segment an activity trace containing a sequence of objects used in a number of consecutive and non-interleaving activities. To do this, they proposed an activity recognition algorithm, *KeyExtract*, which uses the list of discriminatory key objects from all activities to identify the activities present in a trace. They further proposed two heuristic segmentation algorithms, *MaxGap* and *MaxGain*, to detect the boundary between each pair of activities identified by *KeyExtract*. Boundary detection is based on the calculation, aggregation, and comparison of the relative weights of all objects sandwiched in any two key objects representing adjacent activities in a trace. Though the mining-based approach has a number of challenges relating to information retrieval, relation identification and the disambiguation of term meaning, nevertheless, it provides a feasible alternative to model a large amount of activities. Initial research has demonstrated the approach is promising.

Mining-based approaches are similar to data-driven approaches in that they all adopt probabilistic or statistical activity modelling and recognition. But they are different from each other in the way the parameters of the activity models are decided. The mining-based approaches make use of publicly available data sources avoiding the "cold start" problem. Nevertheless, they are weak in dealing with the idiosyncrasies of activities. On other hand, data-driven approaches have the strength of generating personalized activity models, but they suffer from issues such as "cold start" and model reusability for different users.

2.4.2 Logic-Based Approach

The rationale of logical approaches is to exploit logical knowledge representation for activity and sensor data modelling, and to use logical reasoning to perform activity recognition. The general procedure of a logical approach includes (1) to use a logical formalism to explicitly define and describe a library of activity models for all possible activities in a domain, (2) to aggregate and transform sensor data into logical terms and formula, and (3) to perform logical reasoning, e.g., deduction, abduction and subsumption, to extract a minimal set of covering models of interpretation from the activity model library based on a set of observed actions, which could explain the observations.

Even though each task can be undertaken in different ways the role of each task is specific and unique. Normally, the first step is to carry out knowledge acquisition, which involves eliciting knowledge from various knowledge sources such as domain experts and activity manuals. The second step is to use various knowledge modelling techniques and tools to build reusable activity structures. This will be followed by a domain formalization process in which all entities, events and temporal and spatial states pertaining to activities, along with axioms and rules, are formally specified and represented using representation formalism. This process usually generates the domain theory. The following step will be the development of a reasoning engine in terms of knowledge representation formalisms to support the inference. In addition, a number of supportive system components will be developed, which are responsible for aggregating and transforming sensor data into logical terms and formula. With all functional components in place, activity recognition proceeds by passing the logical representation of sensor data onto the reasoning engine. The engine performs logical reasoning, e.g., deduction, abduction or induction, against the domain theory. The reasoning will extract a minimal set of covering models of interpretation from the activity models based on a set of observed actions, which could semantically explain the observations.

There exist a number of logical modelling methods and reasoning algorithms in terms of logical theories and representation formalisms. One thread of work is to map activity recognition to the plan recognition problem in the well-studied artificial intelligence field [66]. The problem of plan recognition can be stated in simple terms as: given a sequence of actions performed by an actor, how to infer the goal pursued by the actor and also to organize the action sequence in terms of a plan structure. Kautz et al. [67] adopted first-order axioms to build a library of hierarchical plans. They proposed a set of hypotheses such as exhaustiveness, disjointedness and minimum cardinality to extract a minimal covering model of interpretation from the hierarchy, based on a set of observed actions. Wobke [68] extends Kautz's work using situation theory to address the different probabilities of inferred plans by defining a partial order relation between plans in terms of levels of plausibility. Bouchard et al. [69] borrow the idea of plan recognition and apply it to activity recognition. They use action Description Logic (DL) to formalize actions and entities and variable states in a smart home to create a domain theory. They model a plan as a sequence of actions and represent it as a lattice structure, which, together with the domain theory, provides an interpretation model for activity recognition. As such, given a sequence of action observations, activity recognition amounts to reasoning against the interpretation model to classify the actions through a lattice structure. It was claimed that the proposed DL models can organize the result of the recognition process into a structured interpretation model in the form of a lattice, rather than a simple disjunction of possible plans without any classification. This minimizes the uncertainty related to the observed actor's activity by bounding the plausible plans set.

Another thread of work is to adopt the highly developed logical theory of actions, such as the Event Calculus (EC) [70], for activity recognition and assistance. The EC formalizes a domain using fluents, events and predicates. Fluents are any properties of the domain that can change over time. Events are the fundamental instrument of change. All changes to a domain are the result of named events. Predicates define relations between events and fluents that specify what happens when and which fluents hold at what times. Predicates also describe the initial situation and the effects of events. Chen et al. [71] proposed an EC-based framework in which sensor activations are modelled as events, and object states as properties. In addition, they developed a set of high-level logical constructors to model compound activities, i.e. the activities consisting of a number of sequential and/or parallel events. In the framework, an activity trace is simply a sequence of events that happen at different time points. Activity recognition is mapped to deductive reasoning tasks, e.g., temporal projection or explanation, and activity assistance or hazard prevention is mapped to abductive reasoning tasks. The major strength of this work is its capability to address temporal reasoning and the use of compound events to handle uncertainty and flexibility of activity modelling.

Logical activity modelling and reasoning is semantically clear and elegant in computational reasoning. It is also relatively easy to incorporate domain knowledge and heuristics for activity models and data fusion. The weakness of logical approaches is their inability or inherent infeasibility to represent fuzziness and uncertainty. Most of them offer no mechanism for deciding whether one particular model is more effective than another, as long as both of them can be consistent enough to explain the actions observed. There is also a lack of learning ability associated with logic-based methods.

2.4.3 *Ontology-Based Approach*

Using ontologies for activity recognition is a recent endeavour and has gained growing interest. In the vision-based activity recognition community, researchers have realized that symbolic activity definitions based on the manual specification of a set of rules suffer from limitations in their applicability because the definitions are only deployable to the scenarios for which they have been designed. There is a need for a commonly agreed explicit representation of activity definitions or an ontology. Such ontological activity models are independent of algorithmic choices, thus facilitating portability, interoperability and reuse and sharing of both underlying technologies and systems. Chen et al. [72] propose activity ontologies for analysing social interaction in nursing homes, Hakeem et al. [73] for the classification of meeting videos, and Georis et al. [74] for activities in a bank monitoring setting. To consolidate these efforts and to build a common knowledge base of domain ontologies, a collaborative effort has been made to define ontologies for six major domains of video surveillance. This has led to a video event ontology [75] and the corresponding representation language [76]. For instance, Akdemir [77] used the video event ontologies for activity recognition in both bank and car park monitoring scenarios. In principle, these studies use ontologies to provide common terms as building primitives for activity definitions. Activity recognition is performed using individually preferred algorithms, such as rule-based systems [73] and finite-state machines [77].

In the dense sensing-based activity recognition community, ontologies have been utilised to construct reliable activity models. Such models are able to match different object names with a term in an ontology which is related to a particular activity.

For example, a Mug sensor event could be substituted by a *Cup* event in the activity model "*MakeTea*" as *Mug* and *Cup* can both be used for the "*MakeTea*" activity. This is particularly useful to address model incompleteness and multiple representations of terms. Tapia et al. [64] generate a large object ontology based on the functional similarity between objects from WordNet, which can complete mined activity models from the Web with similar objects. Yamada et al. [78] use ontologies to represent objects in an activity space. By exploiting semantic relationships between things, the reported approach can automatically detect possible activities even given a variety of object characteristics including multiple representation and variability. Similar to vision-based activity recognition, these studies mainly use ontologies to provide activity descriptors for activity definitions. Activity recognition can then be performed based on probabilistic and/or statistical reasoning [64, 78].

Ontology-based modelling and representation have been applied to general ambient assisted living. Latfi et al. [79] propose an ontological architecture of a telehealth-based smart home aiming at high-level intelligent applications for elderly persons suffering from loss of cognitive autonomy. Michael et al. [80] developed an ontology-centred design approach to create a reliable and scalable ambient middleware. Chen et al. [81] pioneered the notion of semantic smart homes in an attempt to leverage the full potential of semantic technologies in the entire lifecycle of assistive living i.e. from data modelling, content generation, activity representation, processing techniques and technologies to assist with the provision and deployment. While these endeavours, together with existing work in both vision- and dense sensing-based activity recognition, provide solid technical underpinnings for ontological data, object, sensor modelling and representation, there is a gap between semantic descriptions of events/objects related to activities and semantic reasoning for activity recognition.

Most works use ontologies either as mapping mechanisms for multiple terms of an object [64] or the categorisation of terms [78] or a common conceptual template for data integration, interoperability and reuse [79]. Activity ontologies which provide an explicit conceptualisation of activities and their interrelationships have only recently emerged and have been used for activity recognition. Chen et al. [82] proposed and developed an ontology-based approach to activity recognition. They constructed context and activity ontologies for explicit domain modelling. Sensor activations over a period of time are mapped to individual contextual information and then fused to build a context at any specific time point. They made use of subsumption reasoning to classify the constructed context based on the activity ontologies, thus inferring the ongoing activity. Ye et al. [83] developed an upper activity ontologies that facilitates to the capturing of domain knowledge to link the meaning implicit in elementary information to higher-level information that is of interest to applications. Riboni et al. [84] investigated the use of activity ontologies, in particular, the new feature of rule representation and rule-based reasoning from OWL2, to model, represent and reason complex activities.

2.5 Discussions on Activity Recognition Approaches

This section presents the comparison of different AR approaches and further discusses the relations between activity recognition and other closely related areas. As activity recognition involves a number of research areas, and each area is itself a research topic with considerable literature. The full reviews of these related areas are beyond the scope of this chapter.

2.5.1 Activity Recognition Approach Comparison

Compared with data-driven and mining-based approaches, ontology-based approaches offer several compelling features: Firstly, ontological ADL models can capture and encode rich domain knowledge and heuristics in a machine-understandable and processable way. This enables knowledge based intelligent processing at a higher degree of automation. Secondly, DL-based descriptive reasoning along a timeline can support incremental progressive activity recognition and assistance as an ADL unfolds. The two levels of abstraction in activity modelling, concepts and instances, also allow coarse-grained and fine-grained activity assistance. Thirdly, as the ADL profile of an inhabitant is essentially a set of instances of ADL concepts, it provides an easy and flexible way to capture a user's activity preferences and styles, thus facilitating personalised ADL assistance. Finally, the unified modelling, representation and reasoning for ADL modelling, recognition and assistance makes it natural and straightforward to support the integration and interoperability between contextual information and ADL recognition. This will support systematic coordinated system development by making use of seamless integration and synergy of a wide range of data and technologies.

Compared with logic-based approaches, ontology-based approaches have the same mechanisms for activity modelling and recognition. However, ontology-based approaches are supported by a solid technological infrastructure that has been developed in the semantic web and ontology-based knowledge engineering communities. Technologies, tools and APIs are available to help carry out each task in the ontology-based approach, e.g., ontology editors for context and activity modelling, web ontology languages for activity representation, semantic repository technologies for large-scale semantic data management and various reasoners for activity inference. This gives ontology-based approaches huge advantage in large-scale adoption, application development and system prototyping.

Logic-based approaches are totally different from data-driven approaches in the way activities are modelled and the mechanisms activities are recognised. They do not require pre-existing large-scale dataset, and activity modelling and recognition is semantically clear and elegant in computational reasoning. It is easy to incorporate domain knowledge and heuristics for activity models and data fusion. The weakness of logical approaches is their inability or inherent infeasibility to represent fuzziness and uncertainty even though there are recent works trying to integrate fuzzy logics into the logical approaches. Another drawback is that logical activity models are viewed as one-size-fits-all, inflexible for adaption to different users' activity habits. The logical approach, uses logical formalisms, for example event calculus [71] and lattice theory [85], for representing ADL models and conducts activity explanation and predication through deduction or abduction reasoning. Comparing to the above two data-centric approaches, logical approaches are semantically clear in modelling and representation and elegant in inference and reasoning.

A complete comparison between different approaches in terms of a number of criteria is summarised in Tables 2.1 and 2.2. We have collected the experimental results of these surveyed approaches aiming to establish their performance profiles. Initial findings, which are in line with the findings from [86], have found out that the accuracy of different recognition approaches varies dramatically between datasets. The accuracy also varies between individual activities and is affected by the amount of available data, the quality of the labels that were provided for the data, the number of residents in the space that are interacting and performing activities in parallel, and the consistency of the activities themselves. It becomes apparent that the quantitative comparisons of different approaches will only make sense if the experiments are based on the same activities and sensor datasets. Otherwise, the findings may not be applicable to general cases, and even be misleading.

Cook [86] created a single benchmark dataset that contains eleven separate sensor event datasets collected from seven physical testbeds. Using this dataset, a systematic study has been conducted to compare the performance of three activity recognition models: a naïve Bayes classifier (NBC), a hidden Markov model (HMM), and a conditional random field (CRF) model. The result of recognition accuracy using 3-fold cross validation over the dataset is 74.87, 75.05 and 72.16% for the NBC, HMM and CRF respectively.

Table 2.1 The comparison of data-driven approaches

	Generative	Discriminative
Model type	NB, HMM, LDS, DBNs	NN, SVM, CRF, decision tree
Modelling mechanism	(un)supervised learning from datasets	
Activity recognition method	Probabilistic classification	Similarity or rule-based reasoning
Advantage	Modelling uncertainty, temporal information	Modelling uncertainty, temporal information, Heuristics
Disadvantage	"Cold start" problems, lack of reusability and scalability	

Table 2.2 The comparison of knowledge-driven approaches

	Mining	Logic	Ontology	
Model type	HMM, DBN, SVM, CRF, NN	Logic formula, e.g., plans, lattices, event, trees	HMM, DBN, SVM, CRF, NN	Sensor and activity ontology
Modelling mechanism	Information retrieval and analysis	Formal knowledge modelling	(un)supervised learning from datasets	Ontology engineering
Activity recognition method	Generative or discriminative methods	Logical inference, i.e., deduction, induction	Generative or discriminative methods	Semantic reasoning, e.g., subsumption, consistency
Advantage	No "cold start" problems, using multiple data sources	No "cold start" problems, clear semantics on modelling and inference	Shared terms, interoperability and reusability	No "cold start" problems, multiple models, clear semantics on modelling and inference, interoperability and reusability
Disadvantage	The problems as DDA	Weak in handling uncertainty and scalability	The problems as DDA	Weak in handling uncertainty and time

2.5.2 The Influence of Activity Monitoring on Activity Recognition

The outputs of activity sensing, i.e., sensor data, can affect activity recognition in several aspects. Firstly, in a data driven approach, the sensor type can often drive the selection of an appropriate model. Sensors can yield single or multi-dimensional data (e.g., an accelerometer would be multi-dimensional whereas a temperature sensor would be uni-dimensional), and sensors can either give continuous or discrete measurements. The models need to be modified to fit whatever type of sensor data is being used. At the very least, the variable representing each sensor in a data-driven model must match the sensor type in dimensionality and arity. For example, Oliver et al. [57] use a variety of different sensor types, including audio time-of-arrival, continuous and multi-dimensional computer vision measures, and a set of discrete event from mouse and keyboard, as inputs (observations) of a set of HMMs. Liao et al. [52] use continuous 2-dimensional GPS data as input to a CRF. One solution to adapt activity models to sensor types is to include all available sensors in a discriminative or generative model and allow the model itself to choose the most effective ones for any given situation. This is known as sensor selection or active sensing.

Secondly, the complexity of sensor data will determine to some extent the complexity of activity models. In data-driven approaches, sensor data can be directly fed into the activity models, either generative or discriminative, for model training and/or activity inference. Alternatively, sensor data can be pre-processed, e.g., to reduce the complexity of the data, before they are used in model training and activity inference. There is always a trade-off between the complexity of the sensor data in the model, and the complexity of the model. As a general principle the trade-off is always about reducing the complexity of the model as much as possible without sacrificing representation that is necessary for activity recognition.

For knowledge-driven approaches, sensor data do not directly affect activity models and inference. This is because activity models in knowledge-driven approaches are pre-specified based on domain knowledge rather than driven by sensor data. In addition, in knowledge-driven approaches sensor data are always mapped through pre-processing to the values of properties of the formal activity models. As such, the types and complexity of sensor data will only affect the initial conceptualisation of activity models and the complexity of pre-processing but not the model and inference mechanisms.

2.6 Summary

Activity recognition has become the determinant to the success of the new wave of context-aware personalized applications in a number of emerging computing areas, e.g., pervasive computing and smart environments. Synergistic research in various scientific disciplines, e.g., computer vision, artificial intelligence, sensor networks and wireless communications, has resulted in a diversity of approaches and methods to address this issue. In this chapter we present a survey of the state-of-the-art research on sensor-based activity recognition. We first introduce the rationale, methodology, history and evolution of the approach. Then we reviewed the primary approaches and methods in the fields of activity monitoring, modelling and recognition respectively. In particular we identified key characteristics for each individual field and further derived a classification structure to facilitate systematic analysis of the surveyed work. We have conducted in-depth analysis and comparisons of different methods in each category in terms of their robustness to real-world conditions and real-time performance, e.g., applicability, scalability and reusability. The analysis has led to some valuable insights for activity modelling and recognition.

In addition to the extensive review we have discussed emerging research trends associated with activity recognition. One primary direction is complex activity recognition focusing on the underlying modelling, representation and inference of interleaved, concurrent and parallel activities. The other key direction is to improve reusability, scalability and applicability of existing approaches. Research in this direction has been undertaken in several strands, including multi-level activity modelling, abnormal activity recognition, infrastructure mediated monitoring, and sensor data reuse and repurposing. Another noticeable trend is research on formal activity representation at a higher level of abstraction, e.g., developing dedicated activity representation languages and representing situations and goals. These emerging efforts provide guidance and indication for the future research of activity recognition.

Many research questions have not been touched due to the limited space. For example, we did not elaborate in-depth low-level specific technical issues such as uncertainty, temporal reasoning and sensor data inconsistency. We believe the emerged structure of classification of activity recognition approaches and the comparison of their pros and cons can inform and help interested readers for further exploration

References

1. Mozer MC (1998) The neural network house: an environment that adapts to its inhabitants. In: Proceedings of AAAI spring symposium on intelligent environments
2. Leonhardt U, Magee J (1998) Multi-sensor location tracking. In: Proceedings of the 4th annual ACM/IEEE international conference on mobile computing and networking. ACM, New York, NY, USA, pp 203–214
3. Golding AR, Lesh N (1999) Indoor navigation using a diverse set of cheap, wearable sensors. In: Third international symposium on wearable computers digest of papers, pp 29–36
4. Schmidt A, Beigl M, Gellersen HW (1999) There is more to context than location. Comput Graph
5. Randell C, Muller H (2000) Context awareness by analysing accelerometer data. In: Fourth international symposium on wearable computers digest of papers, pp 175–176
6. Gellersen HW, Schmidt A, Beigl M (2002) Multi-sensor context-awareness in mobile devices and smart artifacts. Mob Netw Appl
7. Van Laerhoven K, Aidoo Ka, Lowette S (2001) Real-time analysis of data from many sensors with neural networks. In: Proceedings of 5th international symposium on wearable computer
8. Foerster F, Fahrenberg J (2000) Motion pattern and posture: correctly assessed by calibrated accelerometers. Behav Res Methods Instruments Comput
9. Laerhoven K, Van Cakmakci O (2000) What shall we teach our pants? In: Fourth international symposium on wearable computers, digest of papers, pp 77–83
10. Lee SW, Mase K (2002) Activity and location recognition using wearable sensors. IEEE Pervasive Comput 1(3):24–32
11. Bao L, Intille SS (2004) Activity recognition from user-annotated acceleration data. In: Ferscha A, Mattern F (eds) Pervasive computing. Springer, Berlin, pp 1–17
12. Patterson DJ, Liao L, Fox D, Kautz H (2003) Inferring high-level behavior from low-level sensors. Presented at the 12 October 2003
13. Nugent CD, Mulvenna MD, Hong X, Devlin S (2009) Experiences in the development of a Smart Lab. Int J Biomed Eng Technol 2:319–331
14. Chan M, Estève D, Escriba C, Campo E (2008) A review of smart homes—present state and future challenges. Comput Methods Programs Biomed 91:55–81
15. Programme A, AAL programme - active assisted living programme - ageing well. http://www.aal-europe.eu/
16. Kern N, Schiele B, Junker H, Lukowicz P, Troster G (2002) Wearable sensing to annotate meeting recordings. In: Proceedings - international symposium on wearable computers, ISWC

17. Lukowicz P, Ward JA, Junker H, Stäger M, Tröster G, Atrash A, Starner T (2004) Recognizing workshop activity using body worn microphones and accelerometers. Presented at the 2004
18. Aggarwal JK, Ryoo MS (2011) Human activity analysis: a review. ACM Comput Surv. 43(3):16
19. Ashbrook D, Starner T (2003) Using GPS to learn significant locations and predict movement across multiple users. Pers Ubiquitous Comput. 7(5):275–286
20. Liao L, Patterson DJ, Fox D, Kautz H (2007) Learning and inferring transportation routines. Artif Intell
21. Sung M, DeVaul R, Jimenez S, Gips J, Pentland A (2004) Shiver motion and core body temperature classification for wearable soldier health monitoring systems. In: Eighth international symposium on wearable computers, 2004. ISWC 2004
22. Harm H, Amft O, Roggen D, Tröster G (2008) SMASH: a distributed sensing and processing garment for the classification of upper body postures. In: Proceedings of the 3rd international ICST conference on body area networks
23. Pantelopoulos A, Bourbakis NG (2010) A survey on wearable sensor-based systems for health monitoring and prognosis
24. Dakopoulos D, Bourbakis NG (2010) Wearable obstacle avoidance electronic travel aids for blind: a survey
25. Yoo J, Cho N, Yoo H-J (2008) Analysis of body sensor network using human body as the channel. In: Proceedings of the ICST 3rd international conference on body area networks. ICST (Institute for computer sciences, social-informatics and telecommunications engineering), ICST, Brussels, Belgium, pp 13:1–13:4
26. Cooper RA, Ding D, Simpson R, Fitzgerald SG, Spaeth DM, Guo S, Koontz AM, Cooper R, Kim J, Boninger ML (2005) Virtual reality and computer-enhanced training applied to wheeled mobility: an overview of work in pittsburgh. Assist Technol 17(2):159–170
27. Au LK, Wu WH, Batalin MA, Stathopoulos T, Kaiser WJ (2008) Demonstration of active guidance with SmartCane. In: 2008 international conference on information processing in sensor networks (ipsn 2008), pp 537–538
28. Kim J, He J, Lyons K, Starner T (2007) The gesture watch: a wireless contact-free gesture based wrist interface. In: Proceedings - international symposium on wearable computers, ISWC
29. Madan A, Caneel R (2004) Towards socially-intelligent wearable networks
30. Wang Q, Timmermans A, Chen W, Jia J, Ding L, Xiong L, Rong J, Markopoulos P (2018) Stroke patients' acceptance of a smart garment for supporting upper extremity rehabilitation. IEEE J Transl Eng Heal Med 6:1–9
31. Wilson D, Atkeson C (2005) Simultaneous tracking and activity recognition (STAR) using many anonymous, binary sensors. In: Proceedings of the third international conference on pervasive computing, (PERVASIVE2005)
32. Wren CR, Tapia EM (2006) Toward scalable activity recognition for sensor networks. In: Lecture notes in computer science (including subseries Lecture notes in artificial intelligence and lecture notes in bioinformatics)
33. Srivastava MB, Muntz R, Potkonjak M (2001) Smart kindergarten: sensor-based wireless networks for smart developmental problem-solving enviroments. In: Proceedings of the 7th annual international conference on mobile computing and networking - MobiCom '01 (2001)
34. Hollosi D, Schröder J, Goetze S, Appell JE (2010) Voice activity detection driven acoustic event classification for monitoring in smart homes. In: 2010 3rd international symposium on applied sciences in biomedical and communication technologies, ISABEL 2010
35. Aipperspach R, Cohen E, Canny J (2006) Modeling human behavior from simple sensors in the home. In: Lecture notes in computer science (including subseries Lecture notes in artificial intelligence and lecture notes in bioinformatics)
36. Philipose M, Fishkin KP, Perkowitz M, Patterson DJ, Fox D, Kautz H, Hähnel D (2004) Inferring activities from interactions with objects
37. Fishkin KP, Philipose M, Rea A (2005) Hands-on RFID: wireless wearables for detecting use of objects. In: Proceedings - international symposium on wearable computers, ISWC
38. Patterson DJ, Fox D, Kautz H, Philipose M (2005) Fine-grained activity recognition by aggregating abstract object usage. In: Proceedings - international symposium on wearable computers, ISWC

39. Hodges MR, Pollack ME (2007) An 'object-use fingerprint': the use of electronic sensors for human identification. In: Proceedings of international conference on ubiquitous computing (UbiComp '07)
40. Buettner M, Prasad R, Philipose M, Wetherall D (2009) Recognizing daily activities with RFID-based sensors. In: Proceedings of the 11th international conference on ubiquitous computing - Ubicomp '09
41. Gu T, Wu Z, Tao X, Pung HK, Lu J (2009) epSICAR: an emerging patterns based approach to sequential, interleaved and concurrent activity recognition. In: 7th annual IEEE international conference on pervasive computing and communications, PerCom 2009
42. Quinn JA, Williams CKI, McIntosh N (2009) Factorial switching linear dynamical systems applied to physiological condition monitoring. IEEE Trans Pattern Anal Mach Intell
43. Horvitz EJ, Breese JS, Heckerman D, Hovel D, Rommelse K (2013) The Lumiere project: Bayesian user modeling for inferring the goals and needs of software users
44. Kautz H, Fox D, Etzioni O, Borriello G, Arnstein L (2002) An overview of the assisted cognition project. In: Proceedings of AAAI
45. Kan P, Huq R, Hoey J, Goetschalckx R, Mihailidis A (2011) The development of an adaptive upper-limb stroke rehabilitation robotic system. J Neuroeng Rehabil 8:33
46. Stikic M, Schiele B (2009) Activity recognition from sparsely labeled data using multi-instance learning. In: Lecture notes in computer science (including subseries Lecture notes in artificial intelligence and lecture notes in bioinformatics)
47. Maurer U, Rowe A, Smailagic A, Siewiorek D (2006) Location and activity recognition using eWatch: a wearable sensor platform. In: Lecture notes in computer science (including subseries Lecture notes in artificial intelligence and lecture notes in bioinformatics)
48. Brdiczka O, Crowley JL, Reignier P (2009) Learning situation models in a smart home. IEEE Trans Syst Man Cybern Part B Cybern 39(1):56–63
49. Chen DT, Yang J, Wactlar H (2005) A study of detecting social interaction with sensors in a nursing home environment
50. Ravi N, Mysore P, Littman ML, Dandekar N (2005) Activity recognition from accelerometer data
51. Vail DL, Veloso MM, Lafferty JD (2007) Conditional random fields for activity recognition. In: Proceedings of the 6th international joint conference on autonomous agents and multiagent systems - AAMAS '07
52. Liao L, Fox D, Kautz H (2007) Hierarchical conditional random fields for GPS-based activity recognition. In: Thrun S, Brooks R, Durrant-Whyte H (eds) Robotics research. Springer, Berlin, pp 487–506
53. Hu DH, Yang Q (2008) CIGAR: concurrent & interleaving goal & activity recognition. In: AAAI conference on artificial intelligence
54. Mahdaviani M, Choudhury T (2007) Fast and scalable training of semi-supervised crfs with application to activity recognition. Adv Neural Inf
55. Guralnik V, Haigh K (2002) Learning models of human behaviour with sequential patterns. In: AAAI workshop on automation as caregiver
56. Modayil J, Levinson R, Harman C (2008) Integrating sensing and cueing for more effective activity reminders. In: AAAI fall symposium AI eldercare new solutions to old problems
57. Oliver N, Garg A, Horvitz E (2004) Layered representations for learning and inferring office activity from multiple sensory channels. Comput Vis Image Underst
58. Pentney W, Philipose M, Bilmes J (2008) Structure learning on large scale common sense statistical models of human state. In: Proceedings of the 23rd national conference on artificial intelligence, vol 3, pp 1389–1395. AAAI Press
59. Wu J, Osuntogun A, Choudhury T, Philipose M, Rehg JM (2007) A scalable approach to activity recognition based on object use. In: Proceedings of the IEEE international conference on computer vision

60. Omar F, Sinn M, Truszkowski J (2010) Comparative analysis of probabilistic models for activity recognition with an instrumented walker. In: Proceedings of the 26th conference on uncertainty in artificial intelligence
61. Sánchez D, Tentori M, Favela J (2008) Activity recognition for the smart hospital. IEEE Intell Syst
62. Wyatt D, Philipose M, Choudhury T (2005) Unsupervised activity recognition using automatically mined common sense. In: Proceedings of 20th national conference on artificial intelligence
63. Perkowitz M, Philipose M, Fishkin K, Patterson DJ (2004) Mining models of human activities from the web. In: Proceedings of the 13th conference on world wide web - WWW '04
64. Tapia EM, Choudhury T, Philipose M (2006) Building reliable activity models using hierarchical shrinkage and mined ontology. In: Lecture notes in computer science (including subseries Lecture notes in artificial intelligence and lecture notes in bioinformatics)
65. Palmes P, Pung HK, Gu T, Xue W, Chen S (2010) Object relevance weight pattern mining for activity recognition and segmentation. Pervasive Mob Comput 6(1):43–57
66. Albrecht D, Zukerman I, Nicholson A (1998) Bayesian models for keyhole plan recognition in an adventure game. User Model User-adapt Interact
67. Kautz Ha (1991) A formal theory of plan recognition and its implementation. Presented at the 1991
68. Wobcke W (2002) Two logical theories of plan recognition. J Log Comput 12(3):371–412
69. Bouchard B, Giroux S, Bouzouane A (2006) A smart home agent for plan recognition of cognitively-impaired patients. J Comput 1(5):53–62
70. Shanahan M (1997) Solving the frame problem: a mathematical investigation of the common sense law of inertia. MIT Press
71. Chen L, Nugent C, Mulvenna M, Finlay D, Hong X, Poland M (2008) A logical framework for behaviour reasoning and assistance in a smart home. Lecture notes in computer science (including subseries Lecture notes in artificial intelligence and lecture notes in bioinformatics)
72. Chen D, Yang J, Wactlar HD (2004) Towards automatic analysis of social interaction patterns in a nursing home environment from video. In: Proceedings of the 6th ACM SIGMM international workshop on multimedia information retrieval - MIR '04
73. Hakeem A, Shah M (2004) Ontology and taxonomy collaborated framework for meeting classification. In: Proceedings - international conference on pattern recognition
74. Georis B (2004) A video interpretation platform applied to bank agency monitoring. In: Intelligent distributed surveillance systems (IDSS-04) (2004)
75. Nevatia R, Hobbs J, Bolles B (2004) An ontology for video event representation. In: IEEE computer society conference on computer vision and pattern recognition workshops
76. François ARJ, Nevatia R, Hobbs J, Bolles RC (2005) VERL: an ontology framework for representing and annotating video events. IEEE Multimed 12(4):76–86
77. Akdemir U, Turaga P, Chellappa R (2008) An ontology based approach for activity recognition from video. In: Proceeding of the 16th ACM international conference on multimedia - MM '08
78. Yamada N, Sakamoto K, Kunito G, Isoda Y, Yamazaki K, Tanaka S (2007) Applying ontology and probabilistic model to human activity recognition from surrounding things. IPSJ Digit Cour
79. Latfi F, Lefebvre B, Descheneaux C (2007) Ontology-based management of the telehealth smart home, dedicated to elderly in loss of cognitive autonomy. In: CEUR workshop proceedings
80. Klein M, Schmidt A, Lauer R (2007) Ontology-centred design of an ambient middleware for assisted living: the case of SOPRANO. Context
81. Chen L, Nugent C, Mulvenna M, Finlay D, Hong X (2009) Semantic smart homes: towards knowledge rich assisted living environments. Stud Comput Intell
82. Chen L, Nugent CD, Wang H (2012) A knowledge-driven approach to activity recognition in smart homes. IEEE Trans Knowl Data Eng 24(6):961–974
83. Ye J, Stevenson G, Dobson S (2011) A top-level ontology for smart environments. Pervasive Mob Comput 7(3):359–378

84. Riboni D, Bettini C (2011) OWL 2 modeling and reasoning with complex human activities. Pervasive Mob Comput 7(3):379–395
85. Preuveneers D, den Bergh J, Wagelaar D, Georges A, Rigole P, Clerckx T, Berbers Y, Coninx K, Jonckers V, De Bosschere K (2004) Towards an extensible context ontology for ambient intelligence. In: Markopoulos P, Eggen B, Aarts E, Crowley JL (eds) Ambient intelligence. Springer, Berlin, pp 148–159
86. Cook DJ (2012) Learning setting-generalized activity models for smart spaces. IEEE Intell Syst 2010(99):1

Chapter 3
An Ontology-Based Approach to Activity Recognition

3.1 Introduction

Using ontologies for activity recognition is a recent endeavour and has gained growing interest. In the vision-based activity recognition community, researchers have realised that symbolic activity definitions based on manual specification of a set of rules suffer from limitations in their applicability, i.e., the definitions are only deployable to the scenarios for which they have been designed. There is a need for an explicit commonly agreed representation of activity definitions, i.e., ontologies, for activities that are independent of algorithmic choices, thus facilitating portability, interoperability and reuse and sharing of both underlying technologies and systems. As such, researchers have proposed ontologies for specific domains of visual surveillance. For example, Chen [1] proposed an ontology for analysing social interaction in nursing homes; Hakeem [2] used ontolgoies for the classification of meeting videos, and Georis [3] for activities in a bank monitoring setting. To consolidate these efforts and to build a common knowledge base of domain ontologies, a collaborative initiative has been made to define ontologies for six domains of video surveillance. This has led to a video event ontology [4] and the corresponding representation language [5]. For instance, Akdemir [6] used the video event ontologies for activity recognition in both bank and car park monitoring scenarios. In principle, these studies use ontologies to provide common terms as building primitives for activity definitions. Activity recognition is performed using individually preferred algorithms, such as rule-based systems [2] and finite-state machines [6].

In the object-based activity recognition community, ontologies have been utilised to construct reliable activity models. Such models are able to match an unknown sensor reading with a word in an ontology which is related to the sensor event. For example, a Mug sensor event could be substituted by a Cup event in the activity model "*MakeTea*" as it uses a *Cup*. This is particularly useful to address model incompleteness and multiple representations of terms. For example, Tapia [7] generated a large object ontology based on the functional similarity between objects from WordNet, which can complete mined activity models from the Web with similar objects.

L. Chen and C. D. Nugent, *Human Activity Recognition and Behaviour Analysis*, https://doi.org/10.1007/978-3-030-19408-6_3

Yamada [8] used ontologies to represent objects in an activity space. By exploiting semantic relationships between things, the reported approach can automatically detect possible activities even given a variety of object characteristics including multiple representation and variability. Similar to vision-based activity recognition, these studies mainly use ontologies to provide activity descriptors for activity definitions. Activity recognition is performed based on probabilistic and/or statistical reasoning [7, 8].

More recently, ontology based modelling and representation have been applied in pervasive computing and in particular Ambient Assisted Living. For example, Latfi [9] proposed an ontological architecture of a telehealth based SH aiming at high-level intelligent applications for elderly persons suffering from loss of cognitive autonomy. Michael et al. [10] developed an ontology-centred design approach to create a reliable and scalable ambient middleware. Chen et al. [11] pioneered the notion of semantic smart homes in an attempt to leverage the full potential of semantic technologies in the entire lifecycle of assistive living i.e. from data modelling, content generation, activity representation, processing techniques and technologies to assist with the provision and deployment. While these endeavours, together with existing work in both vision- and object-based activity recognition, provide solid technical underpinnings for ontological data, object, sensor modelling and representation, there is a gap between semantic descriptions of events/objects related to activities and semantic reasoning for activity recognition. Ontologies are currently used as a mapping mechanism for multiple terms of an object as in [7] or the categorisation of terms as in [8] or a common conceptual template for data integration, interoperability and reuse as in [9–12]. Specifically, there is a lack of activity ontologies, i.e., explicit conceptualisation of activities and their interrelationships.

3.1.1 Application Context: Smart Home Based Assisted Living

With the rising aging population and overstretched healthcare resources, technology-driven healthcare delivery to support independent living has attracted increasing amounts of attention. Within this new paradigm, the concepts of Smart Homes (SH) have recently emerged as a viable mainstream approach to achieving this goal [13]. A SH is a residential home setting augmented with a diversity of multi-modal sensors, actuators and devices along with Information and Communication Technologies (ICT) based services and systems [14]. By monitoring environmental changes and inhabitant's activities, an assistive system in a SH can process perceived sensor data, make timely decisions and take appropriate actions to assist an inhabitant perform activities of daily living (ADL), thus extending the period of time living independently within their own home environment .

Currently there are a number of SH projects being developed for the purpose of proof-of-concept demonstration in addition to the establishment of real living environments [15, 16]. There is a broad range of enabling technologies such as sensor networks, data communications and devices, that provide fragments of the necessary functionality required for the SH [17, 18]. This has led to, on the one hand, increasing capabilities of generating massive amounts of sensor data related to SH environments, inhabitants and events, and on the other hand, the high expectation of providing novel advanced ADL recognition and assistance. Trends in the area of SH-based assistive living are now moving from location or time based reminder systems or emergency oriented reactive alert systems towards context-aware cognitive ADL assistance [19]. Cognitive ADL assistance intends to provide just-in-time activity guidance for elderly people and those suffering from for example cognitive deficiencies such as Alzheimer's disease, in completing their ADLs [20]. To achieve this objective it is necessary to (1) monitor an inhabitant's behavior and their situated environment in real time, (2) dynamically fuse and interpret the multiple modalities of signals and features, (3) infer and recognise behaviours, changes or anomalies, continuously in real-time in a progressive way, and (4) provide assistance to help the inhabitant perform the intended activity based on incrementally accumulated sensor data. There is at present a major gap between the potential of data generation and the aspiration of advanced assistance provision in which context-aware personalized ADL assistance at multiple levels of granularity can be provided whenever needed. The central problem behind this gap is the lack of a novel, yet pragmatic, approach to activity recognition which is truly scalable and can be easily deployed within real living environments.

Activity recognition in a SH is presented with a number of challenges as described in Chap. 1. Current researches on activity recognition have mainly focused on the use of probabilistic and statistical analysis methods, the so-called data-driven approach, for single-user single-activity scenarios [21–24]. In this chapter we present an ontology-based knowledge-driven approach to the processing of multi-source sensor data streams for the purposes of activity recognition. The purpose of our study is not to extend existing data-mining methods to address complex activity scenarios, e.g., interleaved or concurrent activities, sensor noise-caused uncertainty and multi-occupancy, however, to develop an alternative activity recognition paradigm that can address the aforementioned challenges. The approach is motivated by the observations that ADLs are daily routines full of common-sense knowledge providing rich links between the environment, events and activities. In addition, user profiles, i.e., an inhabitant's ADL preferences and their specific ways of performing ADLs, provide prior personal-level details about the ADL itself. Such domain and prior knowledge is valuable in creating ADL models, avoiding the need of large-scale dataset collection and training.

3.2 The Ontology-Based System Architecture

Central to the ontology-based approach is the formal explicit ontological modelling and representation of the SH domain, i.e. SH context and ADLs. Ontological activity modelling provides a description-based modelling method. It models activities as a hierarchy of classes with each class described by a number of properties. As such, the generated activity models are able to capture built-in interrelations between objects and activities irrelevant of the sequence in which these objects are used. Context ontologies serve as a situation model that inter-links multi-source sensor data to build a situation at specific time points. These situations can then be interpreted in terms of the ADL models to infer activities. ADL ontologies can be used as a seed ADL classification model. The seed model can, on the one hand, be directly used to interpret situations for activity recognition, and on the other hand, grow naturally by learning undefined activity patterns from ongoing data mining. With the proposed approach presented in this chapter activity recognition is equivalent to performing subsumption reasoning using dynamically constructed SH situations against ADL descriptions. Activity assistance is reduced to instance checking in ontological reasoning, i.e. to discover the most closely-matched activity profile for each user with regard to the recognised activity and subsequently to use the missing properties for assistance provision.

Figure 3.1 depicts the proposed system architecture for ontology-based activity recognition. Central to the architecture is the ontological modelling and representation of ADLs in the context of smart homes. This not only avoids the fundamental difficulty of obtaining labelled data to learn activity models but also provides an effective way to incorporate domain knowledge into both context and activity models. As can be viewed in Fig. 3.1, the ADL Repositories component consists of ADL models at two levels of abstraction, i.e., ADL ontologies as the generic ADL models and User Profiles as a user's personalised ADL model. A user's ADL profile, in essence, is an instance of the corresponding ADL classes in the ADL ontologies with specific binding in terms of the user's activity preferences. Similarly, the Context Repositories component contains conceptual context models, i.e., the context ontologies, which are used to instantiate sensor observations to create situations at discrete time points. A situation at a specific time point is, on one hand, used for activity recognition in real time. On the other hand, it can be archived over a longer period of time and used to extract ADL patterns or detect changes in the way the ADL is being completed. For example, a user regularly makes tea over a period of three months at a specific time in the day, or the duration for making tea increases over a given period of time.

The components in the left-hand column of Fig. 3.1 denote the physical environment of the SH, which consists of users, sensors, actuators and assistive services. The sensors are responsible for monitoring environmental, event and activity contexts. Assistive Services receive instructions from the ADL Assistive Agent and act on the environment, devices and/or the inhabitant through various actuators. Many research

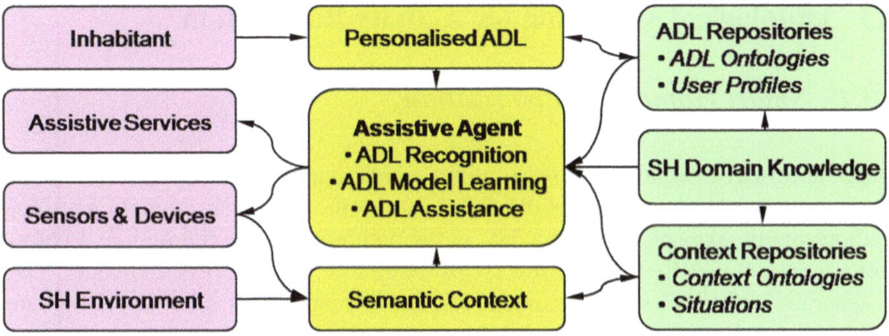

Fig. 3.1 Knowledge-driven activity recognition architecture

issues relating to the hardware aspects in an SH, e.g., optimal deployment of sensors, still remain unsolved, however, are beyond the scope of this chapter.

The Assistive Agent is the core component of the system, which takes as inputs the semantic descriptions of a situation and performs activity recognition with regard to the ADL models in the ADL Repositories. A situation at a specific time is generated by aggregating user-object interactions as described in Sect. 3.3.2. Activity recognition makes use of description-based reasoning capabilities enabled by ontological modelling and representation. The agent performs coarse-grained activity recognition by reasoning semantic situations against generic activity models, i.e., ADL ontologies, and fine-grained activity recognition against a user's ADL profile. This allows an assistive agent to provide generic and personalised assistance respectively in terms of a user's need and the level of details of recognised activities. In addition, given the diversity of ADLs and discrepancy of an individual's living styles and preferences, the manually built ADL ontologies will be treated as the seed of the ADL models. When a large amount of sensor data pertaining to an inhabitant's ADLs are captured and mined over a long period of time, regular activities that have not been modelled before, either a new generic activity or a personalized one with more subtlety can then be identified and extracted. The agent can use these learnt activities to grow ADL ontologies. As such, activity models can evolve, and subsequently improve the performance of activity recognition. The following sections will describe the core constituent components of the system, i.e. ontological activity modelling and recognition, one by one in details.

3.3 Ontological Modelling for Activity Recognition

3.3.1 Smart Home Characterisation

Inhabitants in a SH usually perform routine daily activities in specific circumstances, i.e., in specific locations with specific objects at specific times. For example, brushing teeth normally takes place twice a day, in the bathroom, in the morning and before going to bed. It usually involves the use of toothpaste, a toothbrush and water. This is generally referred to as the context for the corresponding activity. As humans have different lifestyles, habits or abilities, an individual's ADLs and the manner in which they perform them may vary from one person to another. For instance, one person may prefer white coffee and another person may prefer black coffee. Even for the same type of activity, e.g., making white coffee, different people may use different items, e.g., skimmed milk or whole milk, and complete the task in a different order, e.g., adding milk first and then sugar, or vice versa. As such ADLs can be categorized as generic ADLs applicable to all and personalised ADLs taking into account individual subtleties. In addition, ADLs can be conceptualized at different levels of granularity. For example, Grooming can be considered to be comprised of the sub-activities Washing, Brushing and Applying Make-up. There are usually a "*is-a*" and "*part-of*" relationship between a primitive and composite ADLs. All these observations could be viewed as prior domain knowledge that can facilitate activity modelling and recognition.

Ontological modelling is the process to explicitly specify key concepts and their properties for a problem domain. These concepts are organised in a hierarchical structure in terms of their shared properties to form super-class and sub-class relations. For example, *MakeTea* is a subclass of *MakeHotDrink*. Properties establish the interrelations between concepts. For instance, *hasDrinkType* is a property of the *MakeHotDrink* activity that links the *DrinkType* concept (e.g., tea, coffee, chocolate) to the *MakeHotDrink* concept. Both concepts and properties are modelled using the commonly shared terms in the problem community. The resulting ontologies are essentially knowledge models able to encode and represent domain knowledge and heuristics. This avoids the manual class labelling, pre-processing and training processes in the traditional approaches to activity recognition. In addition, ontologies allow software agents to interpret data and reason against ontological contexts, thus enhancing the capabilities of automated data interpretation and inference.

Ontology-based activity recognition approach offers several compelling features: Firstly, ontological ADL models can capture and encode rich domain knowledge and heuristics in a machine-understandable and processable way. This enables knowledge based intelligent processing at a higher degree of automation. Secondly, DL-based descriptive reasoning along a timeline can support incremental progressive activity recognition and assistance as an ADL unfolds. The two levels of abstraction in activity modelling, i.e., concepts and instances, also allow coarse-grained and fine-grained activity assistance. Thirdly, as the ADL profile of an inhabitant is essentially a set of instances of ADL concepts, it provides an easy and flexible way to capture a user's activity preferences and styles, thus facilitating personalised ADL assistance. Finally, the unified modelling, representation and reasoning for ADL modelling, recognition and assistance makes it natural and straightforward to support the integration and interoperability between contextual information and ADL recognition. This will support systematic coordinated system development by making use of seamless integration and synergy of a wide range of data and technologies. In the following sections, we use SH based ambient assisted living to further illustrate these concepts within the realms of ontological activity recognition.

3.3.2 Ontological Context Modelling

SH inhabitants perform ADLs in a diversity of temporal, spatial, environmental contexts. Spatial contexts relate to location information and surrounding entities such as rooms, household furniture and appliances. Event contexts contain background activities and dynamic state changes of appliances and devices. Example events could be the state changes of doors, windows, lights, alarms, a cooker and taps. Environmental contexts are composed of environmental information such as temperature, humidity and general weather conditions. Temporal contexts indicate the time and/or duration. There is a high correlation between ADLs and contexts. For example, a cooking ADL happens in the kitchen with a cooker turned on. A grooming ADL takes place in the bathroom in the morning.

Contextual information is usually captured through various sensors. Each sensor monitors and reflects one facet of a situation. Based on this observation our context modelling is centred on ontological sensor modelling. As can be viewed from the generic conceptual model in Fig. 3.2, sensors are inherently linked to a number of physical and conceptual entities such as objects, locations and states. For example, a contact sensor is attached to a teapot in the second cupboard to the left of the sink in the kitchen. By explicitly capturing and encoding such domain knowledge in a sensor model it is possible to infer the corresponding objects and location from the activation of the sensor. This implies that an inhabitant performs an activity in the inferred location with the inferred object.

As most ADLs require the interlinking and fusion of data from multiple, disparate sensor sources in order to infer the high-level activities, it is necessary to aggregate a sequence of sensor activations to generate a situation at a specific time point. Figure 3.3 shows the situation formation process, which can be described as follows. Each sensor has a default state value for its state property, denoting the state of the object to which it is attached. When a sensor is activated (SA_n), the state property will change. Subsequently, the system will translate the state change as an occurrence of a user-object interaction at the specific time (t_n). As it is difficult, if not impossible, to monitor what happens at detailed levels after an object is interacted with, it is common practice to interpret a sensor activation as a user-object interaction (SSD_n), ignoring how the object is used and when it is de-activated. As such a user-object interaction is equivalent to an instantaneous sensor activation and can be interpreted

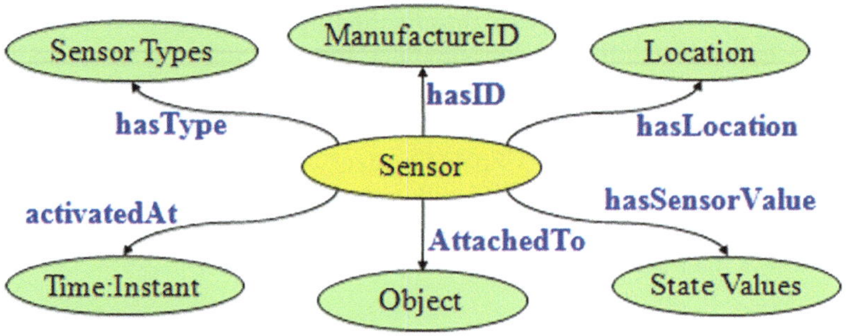

Fig. 3.2 The generic conceptual sensor model

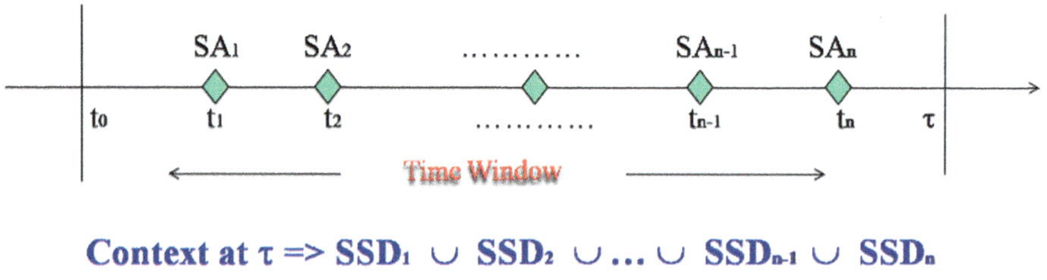

Fig. 3.3 The illustration of the situation formation process

as an object has been used for performing an activity. In this way, by aggregating individual user-object interactions along a timeline (refer to Fig. 3.3) the situation at specific time points can be generated.

The conceptual sensor model depicted in Fig. 3.3 can be formally represented in the SH context ontologies—refer to (a) and (b) in Fig. 3.4. Context ontologies consist of classes and properties for describing SH entities such as *Device, Furniture, Location, Time* and *Sensor*, along with their interrelationships.

To help illustrate our approach we use the *MakeDrink* ADL class hierarchy, as shown in Fig. 3.4, as an example for discussion. An activity model in the ADL ontologies is a concept described by a number of properties that specify relationships between the activity and other entities. Specifically, the *MakeDrink* activity is described with two inherited properties *hasActor* and *hasLocation*, and two specific properties *hasContainer* and *hasAddings*. Its sub-classes, e.g., *MakeHotDrink* and *MakeColdDrink* have additional properties, e.g., drink types. An inhabitant's ADL profile is a set of ADL instances defined by incorporating the user's preferences and life styles. For instance, an inhabitant ADL profile for *MakeDrink* may contain a *MakeHotDrink* instance with properties *hasActor*(*theInhabitant*), *hasLocation*(*kitchen*), *hasContainer*(*cup*), *hasHotDrinkType*(*coffee*) and *hasAddings*(*semi-skimmedMilk*).

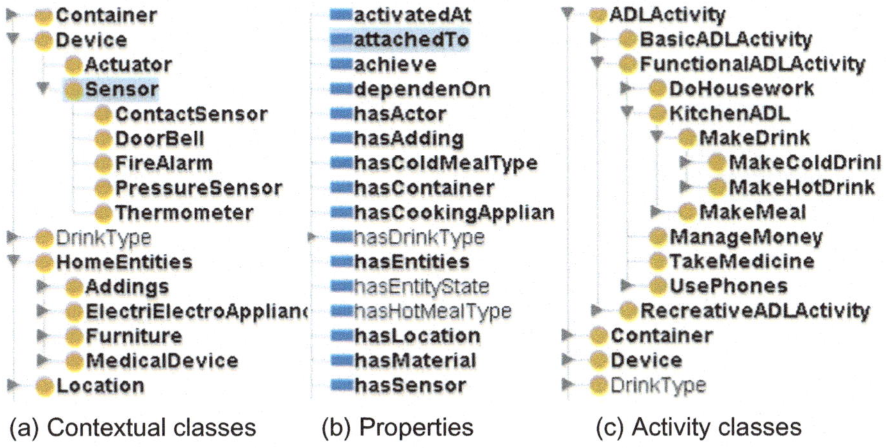

Fig. 3.4 A fragment of the SH domain ontologies

3.3.3 Ontological ADL Modelling

In order to perform activity recognition, computational activity models are required. Based on the nature and characteristics of ADLs we developed a description-based conceptual activity model as presented in Fig. 3.5. In addition to the name and textual description, an activity can be described by a number of properties. These properties relate an activity to other physical items and conceptual entities. They can be categorised into three groups. The first group represents the context, e.g. time, location and actors, within which the activity takes place. The second group represents the causal and/or functional relations, e.g., conditions and effects that are used for inference during high-level activity reasoning. The properties in the third group denote the type and interrelationship between activities. With the diversity of ADLs in a SH this conceptual model will serve as a base model and can be extended to cover ADLs at multiple levels of abstraction.

The conceptual activity model depicted in Fig. 3.5 can be structured and represented in formal ADL ontologies in Fig. 3.4c. The ADL ontology consists of an activity hierarchy in which each node, also called a class, denotes a type of ADL. Each class is described with a number of properties. A property is defined by specifying its domain and range. The domain refers to all classes that can be described by the property and the range refers to all classes whose instances can be assigned to the property. A property describes a class using either a literal or an instance of another class as its value, thus linking two classes. Sub-classes can inherit all properties from its super-class. This can be illustrated using the previous example, i.e., *hasDrinkType* is a property. Its domain is the *MakeHotDrink* activity class and its range the *DrinkType* class. The instances of the *DrinkType* class, e.g., tea and coffee, are the values that the *hasDrinkType* property will take.

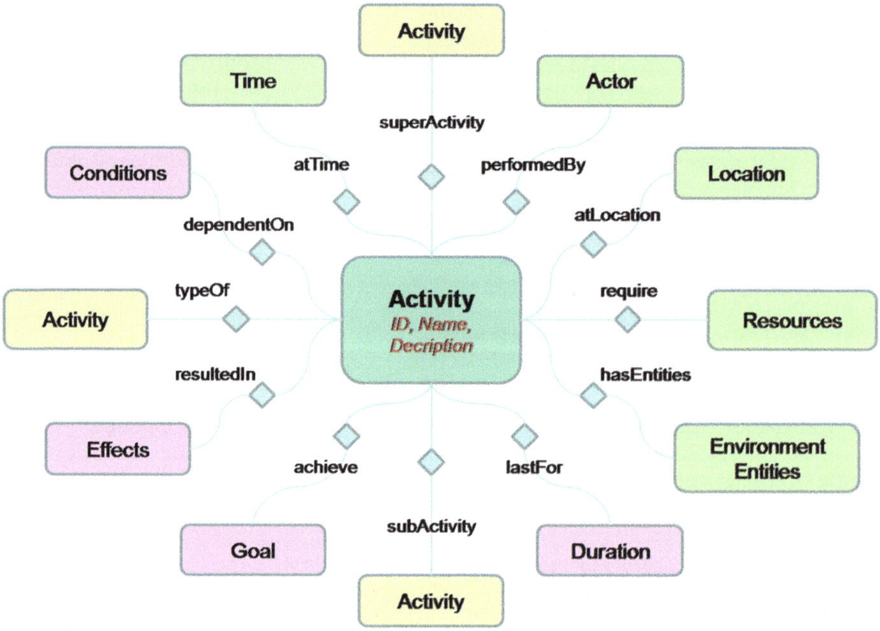

Fig. 3.5 The conceptual activity model

In the knowledge-driven approach, ADL ontologies can be viewed as activity models that establish links between activities and contextual information through activity-based properties. Context ontologies can be regarded as feature models for constructing situations at specific time points that link contextual information with sensor observations through context properties. As such, the whole process of assisted living ranging from low-level sensor data collection, middle-level data fusion, to high-level activity recognition can be streamlined in a unified modelling, representation and reasoning formalism. Activity recognition amounts to constructing situations from sensor observations and reasoning them against activity models.

We carry out knowledge acquisition through interviews, questionnaires and by studying existing documents from which we derive the conceptual models for describing activities and their relations with sensors and objects. Based on SH characterization and the conceptual activity model we develop ADL ontologies using Protégé [25] as shown in Fig. 3.6. The ADL ontology consists of an activity hierarchy in which each node, also called a class, denotes a type of ADL. Each class is described with a number of properties. In a similar way we develop SH context ontologies that consist of classes and properties for describing SH entities such as Device, Furniture, Location, Time and Sensor, and their interrelationships with an activity class. Each sensor monitors and reflects one facet of a situation. By aggregating individual sensor observations, the contextual snapshots at specific time points, or say a situation, can be generated, which can be used to perform activity recognition.

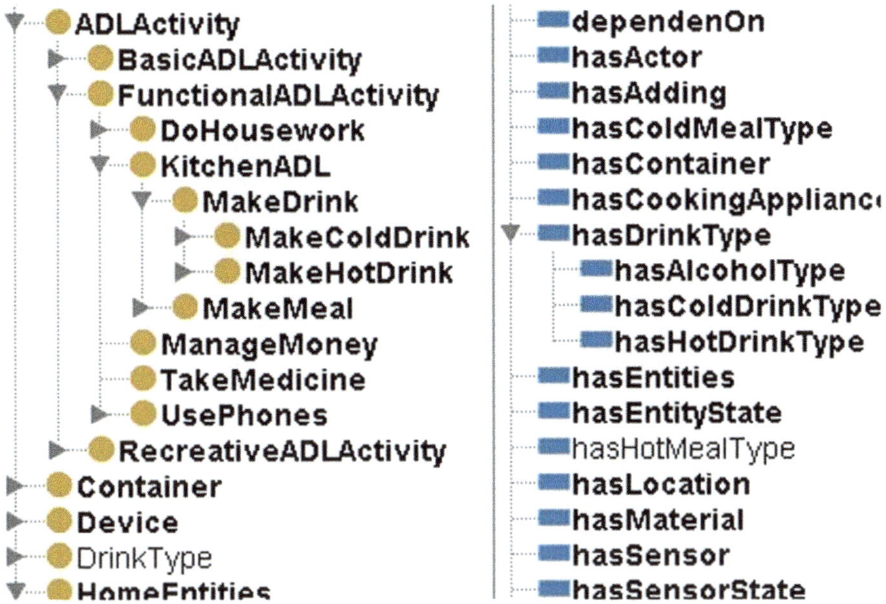

Fig. 3.6 A fragment of the ADL ontologies

Given the nature of sensor data in SH, we develop a two phase semi-automatic approach to generating semantic descriptions. In the first phase, data sources such as sensors and devices are manually semantically described. In the second phase dynamically collected sensor data are first converted to textual descriptors. They are then automatically attached to semantic instances of the corresponding ontological classes to create a semantic knowledge repository. All these operations are performed through demon-like style software tools embedded in the implemented system. the generated semantic data and metadata are archived in a knowledge repository.

3.4 Ontology-Based Mechanisms for Activity Recognition

3.4.1 Theoretical Foundation

The ontological activity modelling and representation is a logical approach in nature in that it uses a Description Logic (DL) based markup language (i.e., OWL and RDF Schema) for specifying conceptual structures and relationships. Both languages support inference and reasoning. The compelling feature of the approach is that it can model domain knowledge at two levels of abstraction. Common, generic activity knowledge can be modelled at the conceptual level as an activity class described by a number of properties. These properties describe the types of objects that can be

Table 3.1 Concept for concept formation syntax and element notations

Formation syntax	Some element notations
$\mathcal{C}, \mathcal{D} \rightarrow \mathcal{A}\mid$ $\leftarrow\leftarrow\leftarrow\leftarrow\leftarrow$ (atomic concept)	$\mathcal{C}1 \sqsubseteq \mathcal{C}2\mid$ (subclass of)
$\top\mid$ $\leftarrow$ (universal concept)	$\mathcal{C}1 \equiv ... \equiv \mathcal{C}$n (equivalent class)
$\perp\mid$ $\leftarrow$ (bottom concept)	$\mathcal{R}1 \sqsubseteq \mathcal{R}$n (subproperty of)
$\neg\mathcal{A}\mid$ $\leftarrow$ (atomic negation)	$\mathcal{R}1 \equiv ... \equiv \mathcal{R}$n (equivalent class)
$\mathcal{C} \cap \mathcal{D}\mid$ $\leftarrow$ (intersection)	$\mathcal{O}1 \equiv ... \equiv \mathcal{O}$n (same individual)
$\forall\mathcal{R}.\mathcal{C}\mid$ $\leftarrow$ (all value restriction)	$\mathcal{C}$i$\sqsubseteq\neg\mathcal{C}$j (disjoint classes)
$\exists\mathcal{R}.\mathcal{C}\mid$ $\leftarrow$ (some value restriction)	$\{\mathcal{O}$i $\neq$ $\mathcal{O}$j$\}$ (different individuals)
$\{\mathcal{O}1, \ldots \mathcal{O}n\}\mid$ (enumeration)	$\mathcal{R}1 \equiv \mathcal{R}2^{-}$ (inverse of)
$\exists\mathcal{R}.\{\mathcal{O}\}\mid$ (property value)	$\mathcal{R}^{+} \sqsubseteq \mathcal{R}$ $\leftarrow$ (transitive)
$\geq$ n$\mathcal{R}.\mathcal{C}$, $\leq$ n$\mathcal{R}.\mathcal{C}$, $=\mathcal{R}.\mathcal{C}\mid$ (min, max, cardinality)	$\mathcal{R} \equiv \mathcal{R}^{-}$ $\leftarrow\leftarrow$(symmetric)

used to perform the activity. In this way, activity models can be created without the requirement of large amounts of observation data and training processes. They are applicable and reusable to a wide range of users.

The core elements of the DL formalism are concepts, roles and individuals. Concepts denote sets of individuals. Roles denote binary relationships between individuals. Individuals are instances of concepts. The vocabulary used for defining concepts and roles of an application domain is referred to as the terminology or the *TBox* in short. All named individuals in terms of vocabulary are referred to as assertions about a real-world domain or the *ABox*. In addition to atomic concepts and roles in the *TBox*, all DL systems allow users to build complex descriptions of concepts and roles. This can be performed using the syntax and constructors of description languages. As DL has a model-theoretic semantics, the statements in the *TBox* and the *ABox* can be interpreted with rules and axioms in DL, thus enabling reasoning and inference, e.g. subsumption and satisfiability reasoning [26].

Consider, in abstract notation, we use the letter $\mathcal{A}$ for atomic concepts, the letter $\mathcal{R}$ for atomic roles, the letter $\mathcal{T}$ for *TBox*, and the letters $\mathcal{C}$ and $\mathcal{D}$ for concept descriptions. Concept descriptions in OWL can be formed using the basic elements in Table 3.1.

DL supports a number of reasoning tasks [26, 27], including satisfiability, subsumption, equivalence, disjointness and consistency. Satisfiability is to check whether a newly defined concept makes sense or whether it is contradictory. The following three tasks are to infer relationships between concepts. For example, subsumption is to find out whether a concept is more general than another one. A concept $\mathcal{C}$ is subsumed by a concept $\mathcal{D}$ if the set of individuals denoted by $\mathcal{C}$ is a subset of the set denoted by $\mathcal{D}$. Actually, all inferences about concept interrelationships can be reduced to satisfiability reasoning.

DL reasoning has been well studied, which supports decidability, completeness and soundness in polynomial time complexity for an inexpressive DL and in exponential time complexity for expressive DLs [27, 28]. With the advanced tableau algorithms and various optimisation techniques, modern reasoners such as FACT++, Pellet and Racer have substantially improved the performance in terms of not only a quantitative change but also a qualitative change from an exponential growth in solution time to an almost constant solution time growth [29]. Current semantic data infrastructures can support semantic classification and queries in seconds against a knowledge base of millions of triples (http://challenge.semanticweb.org). This provides sufficient technological support for our knowledge-driven approach.

With ontological modelling, activities are modelled as activity classes in the ADL ontologies and contextual information such as location and SH objects are modelled as properties for describing activity classes. As such, a situation at a specific time point is actually a concept description created from SH contextual ontologies, denoting an unknown activity. In this case, activity recognition can be mapped to the classification of the unknown activity into the correct position of the class hierarchy of the activity ontologies and the identification of the equivalent activity class. This is the subsumption problem in DL, i.e., to decide if a concept description $\mathcal{C}$ is subsumed by a concept description $\mathcal{D}$, denoted as $\mathcal{C} \sqsubseteq \mathcal{D}$. The commonly used tableau proof system uses negation to reduce subsumption to unsatisfiability of concept descriptions, which can be described below.

- *Reduce subsumption to check unsatisfiability of concept description, i.e., a concept $\mathcal{C}$ is subsumed by a concept $\mathcal{D}$ can be reduced to the checking of satisfiability of concept $\mathcal{C}$ and the negation of concept $\mathcal{D}$, which can be written below*

$$\mathcal{C} \sqsubseteq \mathcal{D} \mapsto \mathcal{C} \cap \neg\mathcal{D}$$

- *Check whether an instance b of this resulting concept description can be constructed* $\leftarrow$
- *Build a tree-like model for the concept description*
- *Transform the concept description in Negation Normal Form*
- *Decompose the description using tableau transformation rules*
- *Stop when a clash occurs or no more rules are applicable*
- *If each branch in the tableau contains a clash, the concept is inconsistent*

Specifically, a situation, i.e., an unknown concept description at a specific time point can be generated by linking sensor observations to properties of the context ontologies (as shown in Fig. 3.3) and incrementally fusing a sequence of sensor observations (as shown in Fig. 3.4). For example, the activation of the contact sensors on a cup and milk bottle can link the cup and milk to the unknown activity through *hasContainer* and *hasAddings* properties. By aggregating sensor observations along a timeline, a specific situation, that corresponds to an unknown activity, could be reached, e.g., *hasTime(10am)*, *hasLocation(kitchen)*, *hasContainer(cup)* and *hasAddings(milk)*. If the closest ADL class in the ADL ontologies that contains as many perceived properties as possible to the situation can be found, e.g., *MakeDrink*, then it can be deemed to be the type of ADL for the identified situation.

3.4.2 Semantic Inference for Activity Recognition

An ontology based knowledge repository $\mathcal{KR}(\mathcal{T}, \mathcal{A})$ as described above consists of a set of terminological axioms $\mathcal{T}$, i.e., ontological concepts and a set of assertional axioms $\mathcal{A}$, i.e., instantiated facts (instances). The activity recognition algorithm is depicted in Fig. 3.7 and described as follows:

a. Detect sensor activations and convert them to corresponding ADL properties in the ontologies.
b. Use context ontologies to aggregate and fuse multiple sensor observations to construct a situation at individual time points.
c. Construct an activity description, denoted as $\mathcal{ATV}$, at two levels of abstraction. $\mathcal{ATV}$-$\mathcal{C}$ denotes the conceptual description of $\mathcal{ATV}$ whereas $\mathcal{ATV}$-I denotes its instance that binds properties with sensor readings.
d. Perform equivalency and subsumption reasoning to check whether $\mathcal{ATV}$-$\mathcal{C}$ is equivalent to any atomic concept $\mathcal{REC}$-$\mathcal{ATV}$ in $\mathcal{T}$. If that is the case, go to step e, otherwise go to step *g*.
e. If $\mathcal{ATV}$-$\mathcal{C}$ is equivalent to an atomic activity concept $\mathcal{REC}$-$\mathcal{ATV}$, then we can recognise $\mathcal{REC}$-$\mathcal{ATV}$ as the type of the underlying activity. We still need to decide whether it is an abstract activity such as *MakeDrink* or a specific activity, e.g., *MakeTea*.
f. Use semantic retrieval to obtain the set of atomic activity concepts $\mathcal{SUB}$-$\mathcal{ATV}$-$\mathcal{SET}$ in $\mathcal{T}$ subsumed by $\mathcal{REC}$-$\mathcal{ATV}$. These are the equivalents and descendants of $\mathcal{REC}$-$\mathcal{ATV}$ in $\mathcal{T}$.

    i. If $\mathcal{SUB}$-$\mathcal{ATV}$-$\mathcal{SET}$ is empty, this means $\mathcal{REC}$-$\mathcal{ATV}$ has no sub-activity concepts and $\mathcal{REC}$-$\mathcal{ATV}$ is a specific activity, e.g. *MakeTea*
    ii. If $\mathcal{SUB}$-$\mathcal{ATV}$-$\mathcal{SET}$ is not empty, this means $\mathcal{REC}$-$\mathcal{ATV}$ has sub-activity concepts and $\mathcal{REC}$-$\mathcal{ATV}$ is an abstract activity, e.g. *MakeDrink*. In this case,

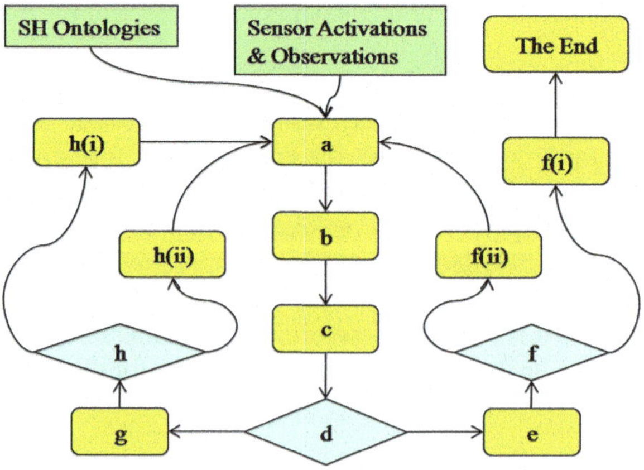

Fig. 3.7 The activity recognition algorithm

a general activity can be recognised such as *MakeDrink*. It will need further sensor data to decide which sub-activity the $\mathcal{ATV}$-$\mathcal{C}$ is, e.g. to decide if it is *MakeColdDrink* or *MakeHotDrink*. This allows incremental activity recognition.

g. If $\mathcal{ATV}$-$\mathcal{C}$ is not equivalent to any atomic activity concept, use semantic retrieval to obtain the most specific atomic concepts $\mathcal{MSC}$-$\mathcal{ATV}$ in $\mathcal{T}$ subsuming $\mathcal{ATV}$-$\mathcal{C}$. In essence, the $\mathcal{MSC}$-$\mathcal{ATV}$ is the direct super-concept of $\mathcal{ATV}$-$\mathcal{C}$.
h. Use semantic retrieval to obtain the set of atomic activity concepts $\mathcal{SUB}$-$\mathcal{ATV}$-$\mathcal{SET}$ in $\mathcal{T}$ subsumed by $\mathcal{MSC}$-$\mathcal{ATV}$. These are the equivalents and descend-ants of $\mathcal{MSC}$-$\mathcal{ATV}$ in $\mathcal{T}$.

   i. If $\mathcal{SUB}$-$\mathcal{ATV}$-$\mathcal{SET}$ is empty, this means $\mathcal{MSC}$-$\mathcal{ATV}$ has no sub-activity concepts and $\mathcal{MSC}$-$\mathcal{ATV}$ is a leaf activity, e.g. *MakeTea*. But the activity has not completed. This means the approach can recognise activities with incomplete sensor data.
   ii. If $\mathcal{SUB}$-$\mathcal{ATV}$-$\mathcal{SET}$ is not empty, the activity recognition process will be similar to the case in *f (ii)* with $\mathcal{MSC}$-$\mathcal{ATV}$ replacing $\mathcal{REC}$-$\mathcal{ATV}$.

It is worth pointing out that the above algorithm describes the process of a single round of activity recognition, i.e., given an activity description (equivalent of a set of sensor data) discover the possible activities. This process could be repeated many times in order to realise continuous progressive activity recognition. To illustrate the algorithm, assume that a *kitchen door* sensor has activated and further the *cup* is only used for making a drink. Suppose that the contact sensor attached to a *cup* is activated. This means that the *cup*, as an instance of *Container*, is used in an ADL. As the *Container* class is the filler of *hasContainer* property, it can be inferred that the *hasContainer* property is assigned the value cup. Since the *hasContainer* property is used to describe the *MakeDrink* class, it can then be inferred that a *MakeDrink* ADL has taken place. Though it is not possible to ascertain whether the ADL is *MakeHotDrink* or *MakeColdDrink* as both ADLs have the *hasContainer* property, nevertheless, based on the limited sensor information the agent can identify the high-level ADLs, i.e., the inhabitant is performing a *MakeDrink* ADL. If, as the ADL unfolds, we obtain sensor data about the use of coffee, then we can determine that the inhabitant is making a hot drink. From the above discussion, it is apparent that the proposed approach can monitor the unfolding of an ADL and continuously recognize the ultimate ADL with an increasing level of accuracy and certainty, which may be considered as not being possible with other approaches.

3.4.3 Real-Time, Continuous Activity Recognition

Activity classes in ADL ontologies are structured in a hierarchical tree with subclasses inheriting all properties from their super-classes. The closer to the leaf of the class tree the more properties with which the activity is described, and the more specific the activity is. When an ADL is performed in the real world along the temporal dimension, the contextual information related to the ADL will be captured incrementally. With less contextual information, e.g., at the initial stage of the ADL, subsumption reasoning can only classify an ADL to a generic activity class. Nevertheless, as the ADL unfolds, more contextual information will become available to enable the creation of an increasingly rich activity description. As such, by performing ADL subsumption reasoning dynamically along a timeline, as shown in Fig. 3.3, it is possible to recognize an ongoing ADL continuously and progressively.

As ontological activity models are description-based, i.e., based on the values or status of their properties, they do not model temporal aspects explicitly in the activity models themselves. Nevertheless, as the status of a property is coupled with sensor activation and a sensor activation is time stamped, temporal data are actually implicitly embedded in the occurrence of a user-object interaction. As such, the handling of the temporal aspects is carried out at the system operational level, which is described in the following paragraphs.

Two methods are used to handle temporal reasoning in order to support real-time, continuous activity recognition. The first is the sliding time window technique for sensor activation fusion. A time window is a fixed duration of time within which all sensor activations are aggregated to generate an activity description. The generated description serves as the input to the subsumption reasoner for activity recognition. If an activity is successfully recognised at a time point within the time window, e.g., the 3rd minute, the algorithm will clear all activated sensors accumulated so far and the time window will re-start from the time point (e.g., the 3rd minute) again, i.e., the time window is sliding each time an activity is recognised. The sliding window technique provides a mechanism for activity partitioning along a timeline and also a method to decide which sensors should be discarded and which should be aggregated to form an activity description. If the algorithm cannot recognise an activity within the time window, the system will deem that the activity is aborted. It will then abort this round of recognition reasoning, i.e., clear all existing activated sensors and re-start sensor monitoring and time window again. Note that within the time window if the system does not receive any user-interactions within a fixed amount of time after the last sensor activation it will provide action reminders for the user—i.e., providing activity assistance. As such, the system only aborts activity recognition after it has attempted to help the user and the user does not respond. A possible scenario is that the system can send an alarm if it infers that the user should do something, however, fails to do so. This is the assistance aspect of the system we do not cover in this chapter.

The length of the sliding window is determined as follows. A fixed maximum amount of time is provided as the default duration of the sliding time window. In addition, the duration of an activity is modelled as a duration property in its ontological model. The value of the duration property can be initially decided based on domain knowledge and later refined through learning. As such, the sliding time window can be adjusted once an activity is initially recognised. For example, a *MakeDrink* activity may set its duration property as 4 min and a *MakeMeal* activity as 15 min respectively. If they are recognised in the continuous progressive activity recognition, the time window can be re-set to the corresponding value. In this way, the correct set of sensor activations can be guaranteed to be used for activity description generation, thus improving recognition accuracy.

The second method for handling temporal reasoning is to determine when a recognition operation is performed, i.e., the activity description from sensor fusion within the time window is fed into the under-lying subsumption reasoner for activity recognition. Activity recognition in assistive living is different from activity recognition in traditional data mining where a prior dataset is available and recognition can be performed offline. In assistive living, an activity must be continuously recognised in real-time to detect anomalies and difficulties during the performance of the user in order to provide just-in-time assistance. This requires the recognition reasoning as described in Sect. 3.4.1 to be performed repeatedly as the activity unfolds. A recognition operation can be initiated in two ways. The first is to specify a fixed time interval (a frequency), i.e., a recognition operation is performed periodically when the time interval elapses. The second way is to use the sensor activation as a trigger, i.e., each time a user-interaction happens (i.e., an action has been performed) the recognition operation is performed. With the ontological activity models, the sensor activation and aggregation models and the two temporal handling methods, the system is able to support real-time continuous activity recognition.

Another feature of the algorithm is that it can offer both coarse-grained and fine-grained activity recognition. The former is based on subsumption reasoning against TBox, i.e., concept descriptions at the terminological level. In this case, following the activity recognition algorithm in Sect. 3.4.1, an extra step may be to compare the properties of the recognised activity with the properties identified by sensor observations. The missing properties can then be used to suggest the next action(s). For example, if the *MakeTea* activity is described by properties *hasContainer*, *hasAddings* and *hasHotDrinkType*, and the sensors observe tea as *HotDrinkType* and cup as Container, then advice on *Addings* such as milk or sugar can be provided.

Fine-grained activity recognition and assistance is based on subsumption reasoning against ABox, i.e., a user's ADL profiles at the instance level. In this case, once the type of activity that an inhabitant performs (as described in Sect. 3.3.2) is recognised, the instances of the inhabitant performing the type of activity in terms of his/her ADL profile can be retrieved from the $\mathcal{KR}(\mathcal{T}, \mathcal{A})$. The discovered ADL instance can then be analysed in terms of sensor observations. Fine-grained activity recognition and assistance will not only recommend the types of action to be performed, but also the items/objects used for the action. For example, suppose an unknown ADL has been identified in the context of *hasContainer(cup), hasAddings(milk)* and *hasHotDrinkType(coffee)*, using the aforementioned recognition mechanism, the *MakeCoffee* ADL can be firstly recognised. Suppose the inhabitant has a special way of making coffee that is modelled in the ABox as *myWayOfMakeCoffee*. Further *myWayOfMakeCoffee* is described by *hasContainer(cup), hasAddings(milk), hasHotDrinkType(coffee), hasAddings(sugar)* and *hasHotWater(hotWater)*. Through comparison, an assistive agent can infer that sugar and hot water are needed in order to complete the ADL. As the matching happens at the instance level, i.e. based on the way the user performs the activity and what actually happened, it can provide users with personalised assistance.

3.5 An Example Case Study

3.5.1 A Prototype System

The proposed approach has been implemented in a feature-rich context-aware assistive system as shown in Fig. 3.8. When the system is in operation within a smart home, it obtains real-time sensor activations from a designated communication port that is connected to an external Tynetec receiver. Each time a sensor is activated, it will aggregate the information with previously collected activated sensors to generate an activity description. The description is then fed to the reasoning engine to infer the potential activity against activity models and profiles. As an actor interacts with the objects in sequence in real time, sensor activations are continuously fed into the system. As such, recognition operations are repeatedly performed to realise continuous progressive activity recognition. As can be seen in Fig. 3.8, the system can dynamically display the activated sensor sequence, the incrementally recognised activities, and system status and data.

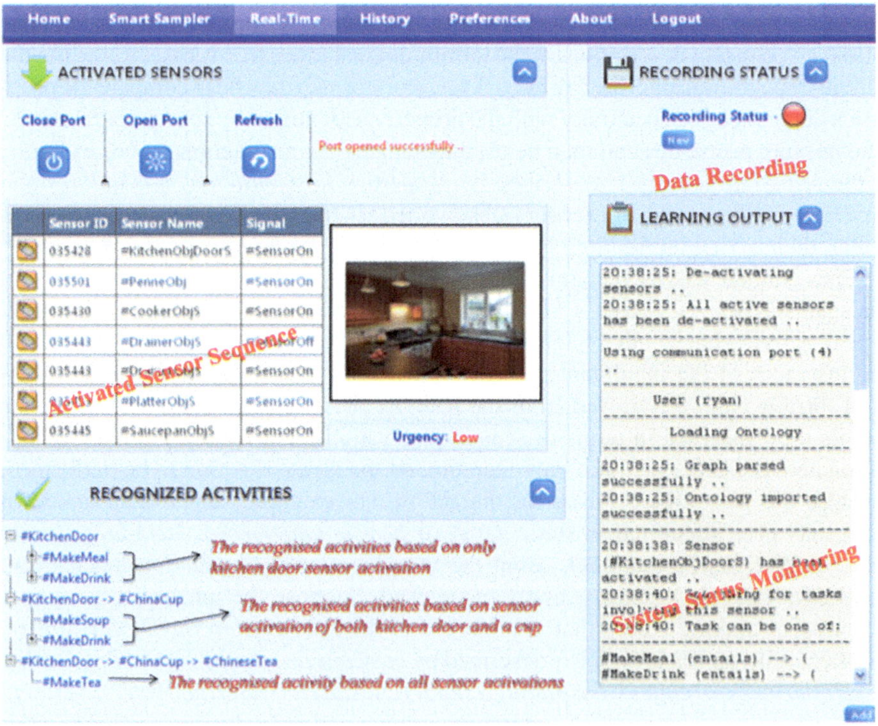

Fig. 3.8 The system interface in real time operation mode

3.5.2 Experiment Setup

To illustrate the use of the methods and algorithms described in previous sections, a feature-rich context-aware assistive system has been developed (Table 3.2). Full detail of the system is elaborated in Chap. 9. and Sect. 9.2. in this case study, eight typical ADLs (refer to Table 3.3) are selected for the purposes of experimentation. For each activity, the required objects for performing the activity were identified and to each of them, an appropriate type of sensor was attached. For example, a tilt sensor was attached to a kettle for detecting the action of pouring water. For each activity, we specify how it is performed based on domain knowledge. It is specified as a sequence of user-object interactions. The interaction is detected by sensors when a user uses the objects to perform the activity. For example, the activity "*making tea*" is specified as the sequence "*go to kitchen, take a cup, take a teabag, add hot water, add milk and add sugar*". These activities and all the sensors are modelled in ontologies and represented in OWL, which are uploaded into the system during system start-up.

In order to test not only functionalities but also scalability and robustness, each activity was designed to be performed in three different ways, leading to three types of activity specification. Table 3.2 shows two selected activities, each with three types of specification. The Type 1 activity specification can be viewed as the "standard" way of performing a specific activity. The experiment using the Type 1 activity specification is mainly aimed to test the functionalities and provide a baseline of activity recognition.

The Type 2 activity specification is obtained by deliberately changing the sequence of the objects being interacted with by a user (refer to Table 3.2). The main objective of this experiment is to test system scalability, i.e., if the system can still recognise the activity even if the way (the sequence of the objects used) of performing the activity has changed. The Type 3 activity specification is obtained by deliberately introducing sensor noise (uncertainty) to the Type 1 specification. We simulate a faulty (fails to

Table 3.2 Two example activity specifications

Activities		Activities specification (sequences of user-object interactions identified by sensors)
Make tea	Type 1	GoToKitchen, GetCup, PourWater, GetMilk, GetSugar
	Type 2	GoToKitchen, GetCup, **PourWater, GetMilk, GetTea,** GetSugar
	Type 3	GoToKitchen, **GetCup***, PourWater, GetMilk, GetSugar
Brush teeth	Type 1	GoToBathroom, RunSink, GetToothbrush, GetToothpaste, GetMouthwash
	Type 2	GoToBathroom, **GetToothbrush, GetToothpaste, RunSink,** GetMouthwash
	Type 3	GoToBathroom, RunSink, GetToothbrush, **getSoap****, GetToothpaste, GetMouthwash

*faulty sensors that do not fire; **false or extra sensor reading

Table 3.3 Recognition results of the 144 activities

Activities	Actor1		Actor2		Actor3		Sum Y/N	
		Exp1	Exp2	Exp1	Exp2	Exp1	Exp2	
Make tea	TP1	Y	Y	Y	Y	Y	Y	6/0
	TP2	Y	Y	Y	Y	N	Y	5/1
	TP3	Y	Y	Y	N	N	N	3/3
Brush teeth	TP1	Y	Y	Y	Y	Y	Y	6/0
	TP2	Y	Y	Y	Y	Y	Y	6/0
	TP3	Y	Y	Y	Y	Y	Y	6/0
Make coffee	TP1	Y	Y	Y	Y	Y	Y	6/0
	TP2	Y	Y	N	Y	N	Y	4/2
	TP3	Y	Y	Y	Y	Y	Y	6/0
Have bath	TP1	Y	Y	Y	Y	Y	Y	6/0
	TP2	Y	Y	Y	Y	Y	Y	6/0
	TP3	Y	Y	Y	Y	Y	Y	6/0
Watch TV	TP1	Y	Y	Y	Y	Y	Y	6/0
	TP2	Y	Y	Y	Y	Y	Y	6/0
	TP3	Y	Y	Y	Y	Y	Y	6/0
Make chocolate	TP1	Y	Y	Y	Y	Y	Y	6/0
	TP2	Y	Y	Y	Y	Y	Y	6/0
	TP3	Y	Y	Y	Y	Y	Y	6/0
Make pasta	TP1	Y	Y	Y	Y	Y	Y	5/1
	TP2	Y	Y	Y	Y	Y	Y	6/0
	TP3	Y	Y	Y	Y	Y	Y	6/0
Wash hands	TP1	Y	Y	Y	Y	Y	Y	6/0
	TP2	Y	Y	Y	Y	Y	Y	6/0
	TP3	Y	Y	Y	Y	Y	Y	6/0
Sum Y/N	All TPs	24/0	24/0	22/2	23/1	20/4	23/1	136/8

activate) sensor by omitting a user-object interaction and a false sensor reading by adding an irrelevant object interaction. The purpose of this experiment is to test system robustness under instances of information uncertainty.

Approximately 40 sensors were deployed in the experiment, including two tilt sensors, two pressure sensors, and the remaining contact sensors. Each sensor was attached to one of the objects involved in these activities. On average five objects (sensors) were used for each activity. The sliding time window was set to five minutes using the system configuration tool. This value is relatively long as in this experiment we do not set the duration properties for these selected activities. The recognition operation is set to be performed each time a sensor is activated, i.e., a user interacts with an object. This is deemed as necessary in order to provide just-in-time assistance in a real assistive living scenario.

We selected three activities out of the eight activities, i.e., making tea, making coffee and making pasta, to test and evaluate fine-grained activity recognition. For each activity, a special way of performing this specific activity, e.g., an activity profile was specified for each actor. For example, actor A had a profile "*ActorA_Preferred_MakeTea*" that specified the specific objects the actor uses for making tea.

Two types of experiment were designed and conducted. In the first experiment, the actor only used the objects specified by the preferred activity model. This experiment aims to test how accurate the system recognises and further provides fine-grained assistance to users according to their preferences. It is assumed that when a user activity preference is available, the system should be able to remind the user, not just the type of action, e.g., adding milk, but also the specific type of object to be used, e.g., adding the semi-skimmed milk. In the second experiment, the actor will perform the activity but replace one of the objects with the same type but a different object. This experiment aims to test how the system reacts to a noisy activity (information uncertainty). In this case, the noise is of the same type but a different object. For example, a user prefers to use *WholeMilk* for making tea, but the actor actually uses the semi-skimmed milk.

3.5.3 Experiment Procedure

Three male actors with ages 25, 35 and 45 took part in the experiments. Each of the participants performed 3 (type) $\times$ 8 (activity) $=$ 24 activities in terms of the activity specification for two rounds. This produced a total of 24 (activity scenarios) $\times$ 2 (rounds) $\times$ 3 (actors) $=$ 144 activities being performed in the experiment.

For each activity, the experiment was carried out as follows: an actor was vocally prompted by a human evaluator to start performing an activity by following the object sequence of the specified activity scenario. The evaluator observed and recorded the actor's action and the recognition results of the system in an experiment data sheet. This provided the ground truth about individual user-object interactions and the activity for later evaluation. The interval between two consecutive actions was set to approximately 30 s. The exact timing of each user-object interaction and the recognition result was recorded by the system logger from which the run-time of each recognition operation can be extracted. In addition to an in-memory buffer for holding sensor activations within a time window, the system provides two automatic recording facilities to store sensor data permanently. The first was to record all sensor activations into a structured XML document representing all user-object interactions during the recording sessions, including sensor information and temporal data. The collected data was mainly used for advanced data processing, e.g., activity model learning and user profile learning. The second method was similar to a server logger. It recorded everything from the start of the system, including sensor activations, the recognition results at each step, error messages, warning messages (e.g., a sensor battery level-low warnings). The recorded information was archived in a text file. The logging function provided a tool for system debugging and results checking.

For the fine-grained activity recognition experiment, two male actors each performed 2 (type) $\times$ 3 (activity) = 6 activities in terms of the activity specifications, producing a total of 6 (activity scenarios) $\times$ 2 (actors) = 12 activities. The procedure is the same as described above.

3.5.4 Results and Discussions

Activity Recognition Accuracy: Table 3.4 shows the recognition results of the 144 activities, where TP refers to the type of activity, Exp1 and Exp2 refer to the two rounds of experiments respectively, Y and N denote the success and failure of activity recognition, and Sum in each row and column refers to the number of Y and N for a particular type of activity and a particular actor respectively. In the analysis, activity recognition is deemed successful the first time an activity is correctly recognised. The rationale is that once an activity is recognised, there is no need for further recognition effort for the present activity until the next activity starts. The activity recognition system will pass the recognised activity to an assistive system, which will subsequently use the standard activity model to help users to complete the activity.

The metric used for evaluation is recognition accuracy, defined as the percentage of the correctly recognised activities against the total activities of a specific experimental scenario. Table 3.4 shows recognition accuracies for a combination of scenarios. As can be seen from Table 3.4, recognition accuracies for type 1, 2 and 3 activity scenarios are 100%, 91.66% and 91.66% respectively with the overall recognition accuracy of 94.44%. The accuracies for type 2 and 3 experiments are lower than the type 1 experiment. This was expected given that type 1 activities are performed according to the way they had been designed in the activity model. If we use type 1 experiments as a benchmark, then the results (91.66%) from type 2 experiments where activities are performed in an order different from that specified in the activity model, and type 3 (91.66%) experiment where various levels of sensor noise are introduced, proves that the approach is scalable, i.e., adaptive to different activity styles, and robust, i.e., resilient to noise (information uncertainty). Given that this

Table 3.4 Activity recognition accuracy (%)

Actor/Activity	Actor 1	Actor 2	Actor 3	Accuracy (Aggr.)
Type 1	100	100	100	100
Type 2	100	93.75	81.25	91.66
Type 3	100	87.5	87.5	91.66
Accuracy (Aggr.)	100	93.75	89.58	Overall: 94.44

is real-time continuous activity recognition with incomplete sensor data, we believe the results are quite impressive.

We also analysed activity recognition accuracies for individual actors. The goal is to evaluate how much different actors affect the recognition performance of the system. In a real-world environment, the system will be used by different users, the consistency of the system performance is therefore important. As can be seen from Table 3.4, the recognition accuracies for the three actors were 100, 93.75 and 89.58% respectively with small variations. This suggests that the system performance is consistent and stable in real world contexts.

Real Time System Performance: Real time continuous activity recognition requires that the recognition operation occurs periodically during the performance of an activity. As such, the time taken for each recognition operation is critical for system responsiveness. We use the Time per Recognition Operation ($TpRO$) as the metric to evaluate the real time system performance. The $TpRO$ is defined as the interval in second(s) from the time a sensor is activated until the time the system produces a recognised activity. As can be seen from Table 3.5, the average $TpRO$ s for each activity ranges from 2.2s to 3.0s with the overall average $TpRO$ roughly as 2.5s. Given the nature of ADLs, i.e., necessary object movement between locations, and the generic characteristics of users of assistive living, i.e., ageing people or cognitively impaired like dementia patients, the $TpRO$ from the experiment proves that the system can efficiently and effectively respond to real world user-object interactions, thus offering real-time continuous recognition.

By analysing $TpRO$ s for different activities, we also test the hypothesis that complex activities involving more user-object interactions will need more time for subsumption classification reasoning, thus higher $TpRO$ to complete recognition operation. Nevertheless, the $TpRO$ results in Table 3.5 disprove this hypothesis since the 7-object activity *Make Pasta* have a $TpRO$ of 2.4s and the 3-object *Wash Hands* had a $TpRO$ of 3.0s. The reason for this counter-intuitive finding may be relevant to other factors of affecting semantic reasoning, e.g., the location of an activity model in the hierarchical tree structure of activity ontologies, and also the number of similar activities, i.e., requiring more effort to distinguish from each other. While further systematic experimentation is required for an in-depth examination of this finding, the generic level and range of $TpRO$ has proved that the system is applicable in real-world scenarios.

Fine-grained Activity Recognition Accuracy: Table 3.6 shows the recognition results of the 12 fine-grained activities. Here activity recognition is deemed as suc-

Table 3.5 The $TpRO$ for all the eight activities performed by actor2 in experiment2

Activity	Make tea	Make coffee	Make chocolate	Make pasta	Wash hands	Brush teeth	Have bath	Watch TV
Objects	5	6	6	7	4	6	5	4
TpROs	2.4	2.2	2.5	2.4	3	3.7	2.6	2.5

Table 3.6 Fine-grained activity recognition results

Activities		Actor1	Actor2
Make tea	TP1	Y	Y
	TP2	Y	Y
Make pasta	TP1	Y	Y
	TP2	Y	Y
Make coffee	TP1	Y	Y
	TP2	Y	Y

cessful if it meets the following two criteria. The first is that the activity should be initially correctly recognised at some stage during the activity performance as a generic activity, i.e., coarse-grained recognition. The second is that once the activity is recognized, the system can discover the user's profile and remind users of the specific objects they could use for performing the activity. Table 3.6 shows that the system achieves 100% recognition accuracy for the type 1 experiment. For type 2 experiment, the success of recognition needs to meet another criterion, i.e., when the same type but different object from the preferred object is used, the system should be able to recommend the preferred object. Again, the system achieves 100% recognition accuracy.

Figure 3.9 shows a fragment of system logs that contain sensor activation and activity recognition traces. The highlighted trace text (by the three circles) indicates that the system recommended using *WholeMilk* for making tea. While the *SkimmedMilk* was actually used, the system still recommended the use of the preferred *WholeMilk*.

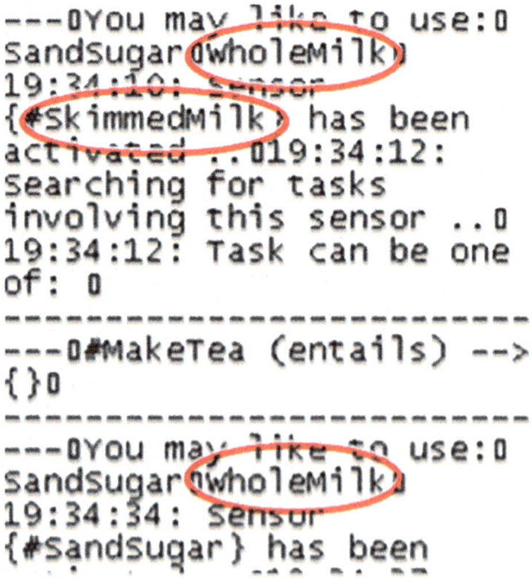

Fig. 3.9 A log trace fragment

Table 3.7 Interaction recognition rate

Sensor types	Total interaction	Captured interaction	Accuracy (%)
Contact	624	611	97.92
Tilt	126	119	94.44
Pressure	36	32	88.89
Sound	18	17	94.44

In a real use case, the system can issue a warning to stop the use of *SkimmedMilk*. This is the feature of assistance provisioning that is not the focus of this chapter.

User-object Interaction Recognition: Throughout the experiment of 144 activities, a total of 804 user-object interactions were performed. We used the User-object Interaction Recognition Accuracy (UoIR), defined as the system's correctly captured interactions against the total occurred interactions, as the metric to evaluate the reliability and performance of the activity monitoring mechanism. This metric takes into account not only unfired or misreads interactions caused by faulty sensors but also those circumstances caused by wireless receivers, communication, and system sampling and conversion mechanisms. As such it is more accurate to reflect the system monitoring performance. Table 3.7 shows the UoIR for different types of sensors with an overall average UoIR of 96.89%. This proves the monitoring and acquisition mechanism of the system as being very reliable.

A detailed examination of the UoIR of individual sensors indicated that the contact sensors had the highest UoIR (97.92%) while the pressure sensor had the lowest UoIR (88.89%). This is consistent with our observation that pressure sensors tend to be very sensitive to the weight variation and the change of sitting posture or other movements can easily generate noise activations.

3.6 Summary

This chapter elaborates a knowledge-driven approach to activity recognition based on ontological modelling and semantic reasoning. We have analysed the nature and characteristics of ADLs upon which argue the necessity of domain knowledge for pattern recognition using multi-source sensor data. Following knowledge engineering practices, we have developed context and ADL ontologies for SH. We have conceived and designed an agent based integrated system architecture to illustrate the realisation of the proposed approach. The compelling feature of the system is the unified ontological modelling and representation for both sensor data and activities, which not only facilitates domain knowledge reuse but also allows the exploitation of semantic reasoning for activity recognition. In particular, a novel activity recognition algorithm is presented that allows continuous activity recognition at multiple levels of abstraction, thus enabling course-grained and fine-grained activity assistance. The mechanism for evolving initial ontological activity models through learning provides a way of marrying the strengths of traditional data-driven approaches with knowledge-driven practices, making our approach flexible, applicable and scalable in terms of rapid system development and deployment.

The implementation of the approach was developed in a feature-rich recognition and assistance system, and a case study has been conducted to demonstrate the use of proposed approach and methods in both real-world and simulated activity scenarios. While the full evaluation of this approach awaits further investigation and user feedback, the initial results have demonstrated the approach is viable and robust in real-world use cases. The approach and the example application scenario have both been developed in a specific application context, namely that of SH-based assistive living. We have not seen any reasons to prevent this approach from being applied to other types of sensorised applications.

The importance of domain knowledge is nowhere more apparent than in the field of assistive living—every individual has their own way of performing activities. We have demonstrated this with respect to one aspect of expertise, namely domain knowledge-driven activity recognition from multi-source sensor data. Future work will focus on large-scale experimentation of a diversity of activity scenarios, multi-user concurrent activity recognition and handling of temporal information such as sequence and duration.

References

1. Chen D, Yang J, Wactlar HD (2004) Towards automatic analysis of social interaction patterns in a nursing home environment from video. In: Proceedings of the 6th ACM SIGMM international workshop on multimedia information retrieval - MIR '04
2. Hakeem A, Shah M (2004) Ontology and taxonomy collaborated framework for meeting classification. In: Proceedings - international conference on pattern recognition
3. Georis B (2004) A video interpretation platform applied to bank agency monitoring. In: Intelligent distributed surveillance systems (IDSS-04)
4. Nevatia R, Hobbs J, Bolles B (2004) An ontology for video event representation. In: IEEE computer society conference on computer vision and pattern recognition workshops
5. François ARJ, Nevatia R, Hobbs J, Bolles RC (2005) VERL: an ontology framework for representing and annotating video events. IEEE Multimed 12:76–86
6. Akdemir U, Turaga P, Chellappa R (2008) An ontology based approach for activity recognition from video. In: Proceeding of the 16th ACM international conference on multimedia - MM '08
7. Tapia EM, Choudhury T, Philipose M (2006) Building reliable activity models using hierarchical shrinkage and mined ontology. In: Lecture notes in computer science (including subseries Lecture notes in artificial intelligence and lecture notes in bioinformatics)
8. Yamada N, Sakamoto K, Kunito G, Isoda Y, Yamazaki K, Tanaka S (2007) Applying ontology and probabilistic model to human activity recognition from surrounding things. IPSJ Digit Cour
9. Latfi F, Lefebvre B, Descheneaux C (2007) Ontology-based management of the telehealth smart home, dedicated to elderly in loss of cognitive autonomy. In: CEUR workshop proceedings
10. Klein M, Schmidt A, Lauer R (2007) Ontology-centred design of an ambient middleware for assisted living: the case of SOPRANO. Context
11. Chen L, Nugent C, Mulvenna M, Finlay D, Hong X (2009) Semantic smart homes: towards knowledge rich assisted living environments. Stud Comput Intell

12. James AB (2014) Activities of daily living and instrumental activities of daily living. In: Willard and spackman's occupational therapy, 12th edn. Wolters Kluwer Health/Lippincott Williams & Wilkins, Philadelphia
13. Chan M, Estève D, Escriba C, Campo E (2008) A review of smart homes—present state and future challenges. Comput Methods Programs Biomed 91:55–81
14. Cook DJ, Hagras H, Callaghan V, Helal A (2009) Making our environments intelligent. Pervasive Mob Comput 5(5):556–557
15. Helal S, Mann W, El-Zabadani H, King J, Kaddoura Y, Jansen E (2005) The gator tech smart house: a programmable pervasive space. Computer (Long Beach, California) 38:50–60
16. Espinilla M, Martinez L, Medina J, Nugent C (2018) The experience of developing the UJAmI Smart Lab. IEEE Access
17. Dundalk Institute of Technology: Home I NetwellCASALA. https://www.netwellcasala.org/
18. Cook DJ, Das SK (2007) How smart are our environments? An updated look at the state of the art. Pervasive Mob Comput 3(2):53–73
19. Chen L, Nugent C, Mulvenna M, Finlay D, Hong X, Poland M (2008) A logical framework for behaviour reasoning and assistance in a smart home. Lecture notes in computer science (including subseries Lecture notes in artificial intelligence and lecture notes in bioinformatics)
20. Philipose M, Fishkin KP, Perkowitz M, Patterson DJ, Fox D, Kautz H, Hähnel D (2004) Inferring activities from interactions with objects. IEEE Pervasive Comput 4(2004):50–57
21. Sánchez D, Tentori M, Favela J (2008) Activity recognition for the smart hospital. IEEE Intell Syst 23(2):50–57
22. Bao L, Intille SS (2004) Activity recognition from user-annotated acceleration data. In: Ferscha A, Mattern F (eds) Pervasive computing. Springer, Berlin, pp 1–17
23. Brdiczka O, Crowley JL, Reignier P (2009) Learning situation models in a smart home. IEEE Trans Syst Man Cybern Part B Cybern
24. Hoey J, Poupart P (2005) Solving POMDPs with continuous or large discrete observation spaces. In: IJCAI international joint conference on artificial intelligence
25. Stanford University: Protégé. https://protege.stanford.edu/
26. Tsarkov D, Horrocks I (2006) FaCT++ description logic reasoner: system description. In: Furbach U, Shankar N (eds) Automated reasoning. Springer, Berlin, pp 292–297
27. Horrocks I, Sattler U (2005) A tableaux decision procedure for SHOIQ. In: IJCAI international joint conference on artificial intelligence
28. Krotzsch M, Simancik F, Horrocks I (2014) Description logics. IEEE Intell Syst
29. Motik B, Shearer R, Horrocks I (2009) Hypertableau reasoning for description logics. J Artif Intell Res

Chapter 4
A Hybrid Approach to Activity Modelling

4.1 Introduction

Activity models play a crucial role in the realization of the SH concept. They are required to support reasoning over real-time streaming sensor data in order to infer the current activity for application-level functions, e.g. to predict the next action or to detect anomalies in ADLs. As a result, the completeness and accuracy of ADL models is critical for an assistive system to function correctly. If an activity is not modelled or the model is not accurate, the activity will not be recognised by an assistive system. Subsequently the system will not be able to provide assistance and/or prediction with regard to this activity.

Modelling ADLs is a challenging task due to their unique characteristics. For example, there are a large number of ADLs in a diversity of categories and they can be modelled at multiple levels of granularity [1]. In addition, most ADLs involve performing a number of actions. The sequence of the actions to be performed is usually dependent on an individual own preference. Furthermore, the manner an ADL is performed is dynamically evolving, such as the change of duration or the order of objects used. This is particularly the case for older people and those suffering from decline of cognitive capabilities.

Currently there are two mainstream approaches to modelling ADLs. One approach, i.e., the data-driven approach, is to learn an individual's activity models from existing behavioural datasets using data mining and machine learning techniques. With this approach activity models are created in two tasks, namely the creation of a probabilistic or statistical activity model, and the training of the model to decide its parameters or mappings [2, 3]. Data-driven approach to ADL modelling has two major drawbacks. The first is the well-known cold-start problem, i.e. requiring a large representative dataset to support model training for each ADL. This problem is exacerbated in the context of assistive living as people are reluctant to disclose their behavioural data due to privacy and ethical considerations. The second drawback is related to model applicability and reusability. Data-driven approach is sensitive to unseen data which makes it difficult to apply the ADL models learnt for

L. Chen and C. D. Nugent, *Human Activity Recognition and Behaviour Analysis*, https://doi.org/10.1007/978-3-030-19408-6_4

one person to another person. This means that with the data-driven approach every activity model for all ADLs for every user needs to be learnt in order to create complete ADL models. Given the large number of ADLs and the cold-start problem, this is a huge challenge, if not impossible, in practice. To mitigate the aforementioned problems, researchers have recently started applying transfer learning techniques to activity modelling and recognition by reusing resources and knowledge, i.e. the source datasets, or features or models, from one user to another in different settings [4–6]. Nevertheless, such research is at its infancy and there are still many open challenges [7].

An alternative to the data-driven approach is to manually define activity models by making use of rich, priori knowledge and domain heuristics. This approach is motivated by the observation that most ADLs are daily routines which normally take place within a specific circumstance of time, location and space with relatively fixed types of objects. Using formal knowledge acquisition and modelling technologies activity models can be created by means of various knowledge modelling tools [8, 9]. As this approach is closely related to knowledge engineering, it is referred to as knowledge-driven approach. Knowledge-driven approach overcomes the cold-start problem and can model activities at multiple levels of abstraction, thus providing the capability to create both generalized and specialized ADL models. For example, ontological activity modelling can model a generic ADL as an ontological activity class and an individual-specific ADL as an instance of the corresponding activity class. Nevertheless, given that ADL models are created manually by domain experts on a case-by-case basis, the approach is questioned about its scalability to generate complete ADL models. In addition, ADL models created by the knowledge-driven approach are perceived as being generic and static. Adapting an individual's ADL models to their changing behaviours is still an open problem.

Rather than trying to reuse resources and knowledge among different users like the transfer learning based research, this chapter introduces an ontology-based hybrid approach by incorporating data-driven learning capabilities into knowledge-driven approach to address the aforementioned problems of activity modelling. The rationale is to provide generic activity models suitable for all users and then create individual activity models through incremental learning. The approach uses semantic technologies as conceptual backbone and technology enablers for ADL modelling, classification and learning. The distinguishable feature of the approach from existing approaches is that ADL modelling is not a one-off effort, instead, a multi-phase iterative process that interleaves knowledge-based model specification and data-driven model learning. The process consists of three key phases. In the first phase the initial seed ADL models are created through ontological engineering by leveraging domain knowledge and heuristics, thus solving the cold-start problem. Ontological activity modelling creates activity models at two levels of abstractions, namely as ontological activity concepts and their instances respectively. Ontological activity concepts represent generic coarse-grained activity models applicable and reusable for all users, thus solving the reusability problem. The seed ADL models are then used in applications for activity recognition at the second phase. In the third phase, the activity classification results from the second phase are analysed to discover new activities and user profiles. These learnt activity patterns are in turn used to update the ADL models, thus solving the incompleteness problem. Once the first phase completes, the remaining two-phase process can iterate many rounds to incrementally evolve ADL models, leading to complete, accurate and up-to-date ADL models.

It is worth pointing out that the research presented in this chapter is based on single-user single-activity scenarios. While complex activity scenarios, e.g. interleaved and concurrent activities, pose many research problems, it is beyond the scope of this chapter. In addition, activity monitoring in this study is based on dense sensing, i.e. one miniaturized sensor is attached to each individual object that are used for performing ADLs. As such, by analysing an inhabitant's interactions with objects of interest it is possible to infer the inhabitant's activity.

4.2 The Hybrid Approach to Activity Modelling

In an attempt to marrying the advantages of both knowledge-based and data-driven approaches, a hybrid approach to activity modelling is developed to incorporate data-driven learning capabilities into knowledge-based framework as described in Chap. 3. The rationale is to provide generic activity models suitable for all users and then create individual activity models through incremental learning. The hybrid approach uses semantic technologies as conceptual backbone and technology enablers for ADL modelling, classification and learning. The distinguishable feature of the hybrid approach from existing approaches is that ADL modelling is not a one-off effort, instead, a multi-phase iterative process that interleaves knowledge-based model specification and data-driven model learning.

The process consists of four key phases depicted in Fig. 4.1. In the phase I, Ontological Activity Modelling, the initial seed ADL models are created through ontological engineering by leveraging domain knowledge and heuristics, thus solving the cold-start problem. Ontological activity modelling creates activity models at two levels of abstractions, namely as ontological activity concepts and their instances respectively. Ontological activity concepts represent generic coarse-grained activity models applicable and reusable for all users, thus solving the reusability problem. For this, ontological knowledge engineering techniques, are utilised to extract and create the initial seed activity models based on domain heuristics and prior knowledge. In phase II, the seed activity models are used as classifiers by activity-based application systems, e.g., an ambient assisted living system, to classify sensor data for the purposes of activity recognition. If an activity has been accurately modelled in the seed activity models, the activity should be recognised. On the other hand, if an activity is not modelled or the model is not accurate, the activity will not be recognised. In the phase III, Activity Learning is performed with the activity classification results from the phase II to discover new activities and user profiles. For this, Data-driven Activity Learning within which data mining-based learning methods are used to learn new activities and a user's activity profile. The learning results from Phase III can then be used to expand or update the seed activity models created in

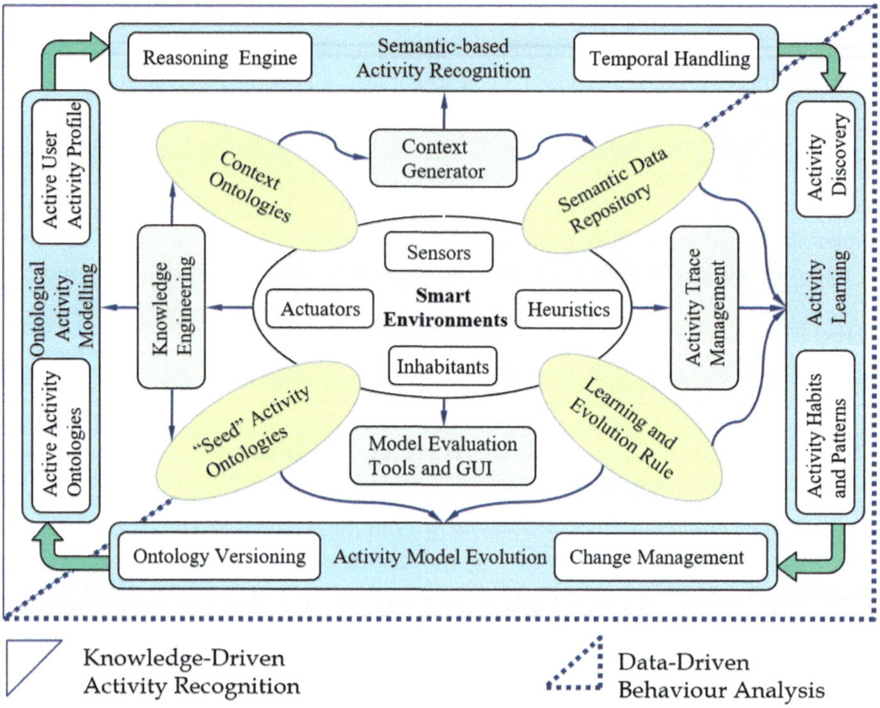

Fig. 4.1 Hybrid framework for activity recognition using knowledge- and data-driven approaches

Phase I. The knowledge evolution or updates to seed ontology is versioned and managed in Phase IV. Thus, Phase IV solves the incompleteness problem in knowledge modelling. The four-phase process can be iterated periodically, thus incrementally evolving and improving the completeness and accuracy of activity models. Among these four phases, Phase I requires human intervention. This includes initial inputs of domain knowledge, manual specification of the seed ontological activity models, and human validation and update of learnt activities at the end of a single iteration. Both Phase II and III are data-driven and completely automatic.

In our previous chapters, we have introduced ontological activity modelling and recognition, the main tasks in Phase I and Phase II. In this chapter, we concentrate on methods and algorithms for learning activity models and user profiles in Phase III. While details of Phase I and II can be found in the aforementioned work, to aid in the understanding of the discussion of the following sections, we briefly outline the rationale and mechanisms of ontology-based activity modelling and recognition.

4.2.1 Ontological Activity Modelling

Ontological activity modelling relates to explicitly specifying activity models using the description logics formalism [10]. It defines an activity as an ontological concept and all actions that are required to perform the activity as the properties of the concept. In addition to action-based properties which are hereafter referred to action properties, an activity model also contains a number of descriptive properties to characterize the manner in which an activity is performed. For example, making tea involves taking a cup from the cupboard, putting a teabag into the cup, adding hot water to the cup, then milk and/or sugar. The ontological model of making tea, i.e., *MakeTea* concept, can be defined by action properties *hasContainer*, *hasTeabag*, *hasHotwater*, *hasMilk*, and *hasFlavor* in conjunction with descriptive properties such as an activity identifier *actID*, start time *actStartTime*, duration *actDuration*, and the sequential order of these objects in performing an activity *actObjSequence*. As action properties are mainly used for defining an activity, they play a crucial role in activity recognition. Descriptive properties, on the other hand, are not determinants in activity recognition. For example, making tea can happen at any time, it can be performed in different sequences and it may take variable amounts of time. Descriptive properties are mainly used to define user's activity profiles, namely to characterize the manner an activity is performed.

Activities can be modelled at different levels of abstraction. As such, ontological activity concepts are usually organized in a hierarchical structure to form superclass and subclass relationships. For example, *MakeTea*, *MakeCoffee*, and *MakeHotChocolate* activities can be modelled as the subclasses of *MakeHotDrink* activity, which is in turn the subclass of *MakeDrink*. Properties establish the relations between ontological activity concepts and the actions required for performing the activities. For example, the *hasContainer* property links the action of preparing a cup to the activity of making tea. Subclasses can inherit properties from *superclasses*. A leaf node of the hierarchy denotes a primitive activity that cannot be further classified. Figure 4.2 presents an excerpt of activity ontologies and associated SH contextual concepts.

Ontological activity concepts define high-level generic activity models which are applicable to anyone. In addition to this, ontological activity modelling can also define the specific way that a person performs an activity, which is usually referred to as user activity profiles. For example, a user always makes English tea at 10 am using skimmed milk and brown sugar. User activity profiles can be defined by creating an instance of a generic ontological activity concept in terms of the user's preference and habits. Ontological activity modelling in Phase I can generate both generic activity models and user activity profiles, thus providing activity models at different levels of abstraction. Aside from activity concepts, other major entities from the domain will also be ontologically modelled. For example, a sensor concept and related properties are developed to establish the relationships between physical sensors, objects, and their locations in addition to the sensor activation time. Further details of these concepts can be found in Chap. 8.

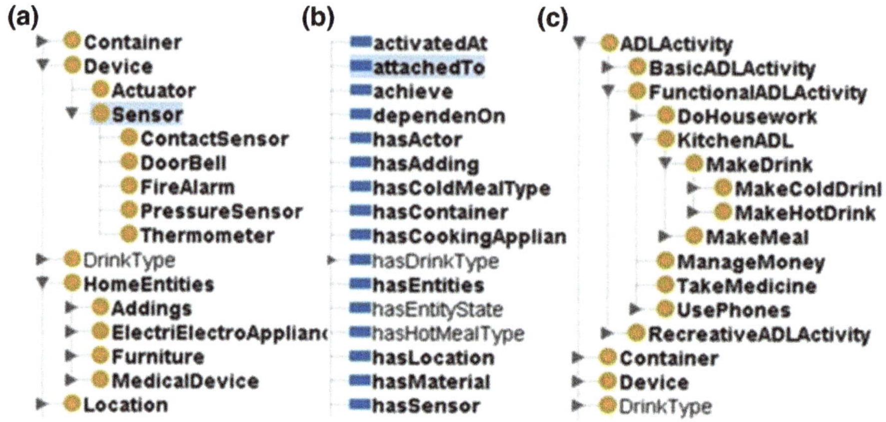

Fig. 4.2 An excerpt of the activity ontologies with **a** entities **b** relationships, and **c** ADLs

4.2.2 Semantics-Based Activity Recognition

In dense sensing based activity monitoring an action of a user interacting with an object is detected through the sensor attached to the object. As such, the activation of a sensor implies that an action has been taken and subsequently an action property relating to the object will be assigned a value. Suppose that a number of sensors are activated along a time line and these sensor observations have been linked to corresponding action properties. At a specific time point, the aggregation of these action properties will create a context denoting an ontological activity description. For example, the activation of the contact sensors on a cup and milk bottle can link the cup and milk to the activity being performed through *hasContainer* and *hasFlavor* properties. Assume that at a specific time, i.e., *hasTime(10 am)*, sensor observations hasLocatio*n(kitchen)*, hasContain*er(cup)*, *hasTeabag* (*English Teabag)*, and *hasFlavor(sand sugar)* are generated, in aggregation, this represents a context for an ongoing activity. If an activity concept in the ADL ontologies, e.g., *MakeTea*, has been defined by this set of action properties, then the activity can be deemed as the type of activity for the perceived context.

The rationale of inferring an activity from sensor observations described above can be formulated as follows: Given a set of action properties instantiated by sensor observations, identify the activity concept in the ADL ontologies that has the same set of action properties. Conceptually, this problem of activity recognition can be mapped to the classification of the activity description using activity ontologies as the classifiers. Technically, the problem can be solved using the subsumption reasoning in description logic, i.e., to decide if a concept description *C* created from sensor observations is subsumed by a concept description *D* within the activity models. Details of the theoretical foundation, reasoning algorithm, and continuous recognition mechanisms for ontology-based activity recognition can be found in Chaps. 2

and 3. It is worth pointing out that the sensor data stream will be first partitioned into segments so that sensor activations within a segment can be aggregated to create an ontological activity description for activity recognition. We have developed a dynamic segmentation model based on the notion of varied time windows for real-time sensor data partition. The model can shrink and expand the window size of segmentation by using temporal information of activity models and sensor data. Further details as discussed in Chap. 7.

To facilitate discussion, we refer to the sequence of sensor observations within a segment as an action trace, i.e., the actions being undertaken within the segment. An action trace is equivalent to a set of action properties in an ontological activity description. With activity models from Phase I and streaming sensor data from applications within a SH, activity recognition in Phase II can produce two types of action traces. If an action trace has a corresponding activity concept in the ADL models, this type of action trace is referred to as a **labelled action trace** or *LAT* in short. Otherwise it is an **unlabelled action trace** or *UAT*. *LATs* can be recognised from the set of action properties while *UATs* cannot be recognised as there are no corresponding activity models in the ADL ontologies or the models are not accurate.

The initial seed activity models generated in Phase I are inevitably incomplete due to the large number of ADLs, the different manner of users performing the ADLs, and the changing user behaviours. As such, when an application within a SH performs activity recognition over a period of time, it will generate large amount of *LATs* and *UATs* that contain information relating to unmodelled activities and the changing behaviours of a user. These action traces can be analysed in Phase III to learn new activities and user activity profiles. New activities increase the completeness of activity models while user activity profiles improve the accuracy of activity models. Sections 4.3 and 4.4 describe the details of the learning mechanisms and methods for Phase III.

4.3 Learning Unmodelled Activities

Activity learning aims to discover the activities that a user performs, however, which have not been modelled in the seed activity ontologies. As there are no models for these activities, they will not be recognised by the activity recognition process in Phase II. Subsequently, they are classified as unlabelled action traces, i.e., *UATs*. The essence of the activity learning is therefore to extract regular activities from *UATs* so that they can be modelled to improve activity models.

We have developed a three-step learning method for this purpose. In the first step, a semantic similarity metric is defined to measure the semantic similarity between two *UATs*. Based on this, an algorithm is then developed to compute the semantic similarity. In the second step, the semantic similarity between any individual *UAT* among all *UATs* is calculated. Based on the similarity metrics, all UATs are classified into a number of subsets where each subset contains semantically similar *UATs*. In essence, each subset corresponds to one unmodelled activity and the number of

UATs within each subset is the number of occurrences of the unmodelled activity during activity monitoring. Given that an unmodelled activity could be a one-off or random behaviour, it would be necessary to determine which discovered activities are regular activities and should be formally modelled. In the third step, the frequency of the occurrence of these discovered unmodelled activities are calculated, and a threshold is specified based on domain heuristics. If the occurrence frequency of an unmodelled activity is equal or greater than the threshold, then the activity will be formally modelled to update the ADL models.

As previously discussed, central to the activity learning method is the definition and computation of semantic similarity between UATs. We define $sim_{uat}(UAT_i, UAT_j)$ as the semantic similarity measure between two *UATs* and denote each UAT as a set of action property-value pairs represented as follows:

$$\begin{aligned} UAT_i &= \{prop_1 - value_1, prop_2 - value_2, \ldots prop_k - value_k\} \\ UAT_j &= \{prop_1 - value_1, prop_2 - value_2, \ldots prop_n - value_n\}. \end{aligned} \tag{4.1}$$

By semantic similarity, we refer to the similarity of two *UATs* in terms of the types of the property values rather than the values themselves. This is because in ontological activity modelling, an activity model is defined by the types of object rather than the objects themselves. For example, the *MakeTea* activity is specified by *hasContainer(x), hasTeabag(y)* . . . and *hasFlavor(z)*. It is the types of the property values, rather than the specific *x, y*, or *z* objects that define the activity. The value of a property, e.g., *x, y*, or *z*, can be any object, e.g., *cup1* or *cup2* for *x*. *English tea* or *Indian tea* for *y*, *white sugar* or *brown sugar* for *z*. As such, the types of objects are the key discriminants to decide if two *UATs* refer to the same type of activity.

We have developed a method to compute the semantic similarity of two *UATs* in terms of the similarity of the two sets of property-value pairs in the two *UATs*. The method works as follows. We first map the set of action properties in a *UAT* to a corresponding set of objects and then derive the corresponding object type for each individual object. Both mappings from action properties to objects and from objects to object types are conducted by recursively unfolding the semantic relations based on ontological relationships modelled in the ADL ontologies. As a result of these mappings, a *UAT* can be transformed into a description of a set of object types, as denoted in the formula below

$$\begin{aligned} UAT_i = \{&objectType\ of\ prop_1 - value_1, \\ &objectType\ of\ prop_2 - value_2, \ldots \\ &objectType\ of\ prop_k - value_k\}. \end{aligned} \tag{4.2}$$

Following this semantic explication and transformation, the similarity of two sets of property-value pairs is equal to the similarity of two sets of object types. As each object type is modelled as a concept in the ADL ontologies, the semantic similarity between two object types (concepts) can be computed based on the signatures of the object concepts. Specifically, the similarity measure can be calculated using the

Jaccard coefficient [11] which is the ratio of the number of shared elements from the intersection of the two sample sets to the number of total elements from the union of the two sets. This is represented as follows:

$$sim_{uat}(UAT_i, UAT_j) = (|UAT_{iot} \cap UAT_{jot}|/|UAT_{iot} \cup UAT_{jot}|. \quad (4.3)$$

Here, UAT_{iot} and UAT_{jot} refer to the set of object types in UAT_i and UAT_j, respectively.

Example 4.1 illustrates the transformation of two *UAT s* and their semantic similarity. Even though the order and specific objects used for each activity is different, the semantic similarity measure equals one, indicating they refer to the same type of activities.

Example 4.1 *UATs are in the form of property-value pairs.*

$$UAT_i \left\{ \begin{array}{c} hasContainer(mug_a), hasTeabag(Englishteabag), hasFlavor(brownsugar), \\ hasHotwater(kettle_a), hasMilk(semiskimmedmilk) \end{array} \right\}$$

$$UAT_j \left\{ \begin{array}{c} hasContainer(mug_a), hasTeabag(Indiateabag), hasFlavor(whitesugar), \\ hasHotwater(kettle_a), hasMilk(skimmedmilk) \end{array} \right\}$$

UATs are in the form of sets of objects.
$\mathrm{UAT_i}$ = *{mug$_a$, English teabag, brown sugar, kettle$_a$, semiskimmed milk}*
$\mathrm{UAT_j}$ = *{mug$_b$, India teabag, kettle$_a$, skimmed milk, white sugar}*
UATs are in the form of sets of object types.
$\mathrm{UAT_i}$ = *{Container, Tea, Sugar, Kettle, Milk}*
$\mathrm{UAT_j}$ = *{Container, Tea, Kettle, Milk, Sugar}*
sim_{uat} *(UATi, UATj)* = 1.

We can compute the semantic similarity of any two *UATs* using the described method and then use the resulting similarity metrics to classify all *UATs* into a number of subsets of *UATs*. For each subset, the semantic similarity $\mathrm{sim_{uat}}$ between any two *UATs* is equal to 1, hence each subset denotes a specific type of activity.

Once distinct activities are discovered through semantic classification, it is necessary to decide whether they are regular activities, random or one-off activities. To this end, we use the daily frequency of occurrence of a *UAT* as the significance measure for the activity it represents. For example, if the daily frequency of occurrence of UAT_k is n, this means the activity UAT_k occurs on average n times a day during the period of monitoring, e.g., once a day for n = 1, twice a day for n = 2, and once every two days for n = 0.5. A threshold value can then be specified for the daily frequency of occurrence based on domain knowledge and heuristics. For example, given that most ADLs are performed on a daily basis, we can reasonably set 0.5 as the threshold value, namely a *UAT* happening once every two days can be regarded as a regular activity. If the daily frequency of occurrence of a *UAT* is greater or equal to the threshold value, the *UAT* can be formally designated as a regular activity. Subsequently, this activity will be modelled to update the activity models. Table 4.1

Table 4.1 The algorithm for learning unmodelled activities

Variables	Descriptions
SU	*the whole set of UATs*
SSU_i	*the ith subset of UATs within which all UATs are semantically similar*
fo_{uat}	*the daily frequency of occurrence of an UAT*
T_{fo}	*the threshold value specified for fo_{uat}*
D	*the duration of activity monitoring in days*

```
set SU, D, T_fo from Phase II outputs
for any UAT_i, UAT_i∈SU, do
  semantic unfolding and transformation as illustrated in Example 1
enddo
set a counter actNum = 0, representing the number of new activities
while ( |SU| > 0)
set UAT_base to an arbitrary member of SU
    create a new subset SSU_actNum with UAT_base as the only member
    for ( 1 ≤ i ≤ |SU| )
        calculate sim_uat(UAT_base, UAT_i), where UAT_i∈SU
        if ( sim_uat(UAT_base, UAT_i) = 1 )
            put UAT_i into the set SSU_actNum
            remove UAT_i from SU
        else
            leave UAT_i in SU
        endif
    endfor
    Increase the counter actNum = actNum + 1
endwhile // this will create actNum subsets SSU_i
for ( 1 ≤ i ≤ actNum )
    calculate fo_uat(UAT, UAT ∈ SSU_i) = |SSU_i | / D
    if (fo_uat(UAT) ≥ T_fo ) recommend to an expert
        SSU_i represents a regular activity
    else
        SSU_i represents a random / one-off activity
    endif
Endfor
```

summarizes the variables, their descriptions and the pseudo code of the algorithm for the presented activity learning method.

4.4 Learning User Activity Profiles

An activity can be performed in many different ways, e.g., using different items of the same object types, in different sequence of actions, at different times and within variable durations. A user activity profile is referred to the specific way of a user performing activities which is the key to personalised assistance in assistive living. To formally specify a user activity profile, we use three attributes, namely an object pattern, duration, and an activity pattern, to characterize the manner that an activity is performed. An object pattern refers to the unique order of objects that an activity is performed whilst an activity pattern describes the frequency and regularity of an activity occurrence, including the starting time(s). Ontological activity modelling can model an activity profile as an instance of the corresponding generic activity concept. Nevertheless, the initial seed activity models do not contain user profile models. This is because the model of a user activity profile is user specific, it can only be defined once a user is identified. In addition, a user's behaviour can change due to physical or mental conditions, thus leading to the change of activity profiles. As such, learning user behaviour from their activity observations is an effective way to create user profiles. An *LAT* represents an activity that has been modelled in the ADL ontologies and recognised in Phase II. Each *LAT* has a corresponding activity label and a sequence of sensor observations denoting the specific undertaking of the activity. Over time for each activity, there will be a set of accumulated *LATs*, which provide a valuable source for user profile discovery. In the following sections, we describe the processes and methods of learning user profiles from real time observations of activity performance, i.e., the *LATs* generated in Phase II.

4.4.1 Object Patterns Detection

We have developed a three-step learning method to discover whether or not a user follows a unique object pattern in performing an activity. In the first step, we define a similarity measure $sim_{lat}(LAT_i, LAT_j)$ in terms of object sequences and develop an algorithm to calculate the similarity of two *LATs*. In the second step, we compute the similarity among all *LATs* of a specific activity and based on the similarity measures, a classification algorithm is developed to classify the set of *LATs* into subsets of *LATs* of the same object pattern. In the third step, we calculate the distribution of frequency of occurrences of all object patterns for the specific activity. The dominant object pattern can then be used to characterize the user activity profile for the specific activity.

Similar to a *UAT*, an *LAT* can be denoted as a set of action property-value pairs, i.e., $LATi = \{prop_1 - value_1, prop_2 - value_2 \ldots prop_k - value_k\}$. We define $sim_{lat}(LAT_i, LAT_j)$ as the similarity measure in terms of object sequences of the two *LATs*. To calculate the similarity measure, we transform an *LAT* from a sequence of action property-value pairs to a sequence of objects through semantic unfolding of ontological concepts. The resulting *LAT* can be represented as a sequence of objects, i.e., $LAT = \{object_1 ofprop_1 - value_1, object_2 ofprop_2 - value_2 \ldots object_k ofprop_k - value_k\}$, where each element $object_i$ is a specific object denoted by its signature. After this transformation, an *LAT* can be treated as an object signature vector, and the similarity of two *LATs* is essentially the similarity between two vectors in a high dimensional space. This can be computed using the generic cosine similarity algorithm [12], as formulated in the Eq. 4.4:

$$sim_{lat}(LAT_i, LAT_j) = (LAT_i, LAT_j)/(||LAT_i||||LAT_j||)$$
$$= \sum_{k=0}^{n}(LAT_{ik} \times LAT_{jk})/\left(\sqrt{\sum_{k=0}^{n} LAT_{ik}}\right)^2 \times \sqrt{\sum_{k=0}^{n}(LAT_{ik})^2} \quad (4.4)$$

The numerator is the dot product of the two *LAT* vectors and the denominator is the product of the magnitudes of the two vectors. i and j are an *LAT* respectively, $i \neq j$, and n is the total number of *LAT s*. A value in the range $[-1, 1]$ can be generated, where -1 signifies the exact opposite object pattern and 1 signifies exactly the same pattern.

In order to make use of the cosine similarity algorithm to compute similarity of *LATs*, we convert the text notation of the elements of an *LAT* to numerical values by allocating each object an object identifier number. The object identifier numbers do not have any meaning, they are simply used to facilitate the similarity computation based on object sequences. Example 4.2 below illustrates three *LATs*, their object signatures, corresponding exemplar object identifier numbers and the similarity measures between them.

Example 4.2 LAT_1 *{*mug_a(1)*, teabag* (2)*, hotwater*(3)*, sand sugar*(4)*, skimmed milk*(5)*}*

LAT_2 *{*mug_b (9)*,teabag*(2)*, whole milk*(8)*, hotwater*(3)*, sand sugar*(4)*}*

LAT_3 *{*mug_a (1)*,teabag*(2)*, hotwater*(3)*, sand sugar*(4)*, skimmed milk*(5)*}*

sim_{lat} $(LAT_1, LAT_2) = 0.7053$

sim_{lat} $(LAT_1, LAT_3) = 1.$

As shown in the above example, LAT_1 and LAT_3 will be classified into the same subset because they follow the same object sequences. Similarly, we can compute the similarity measures for all *LATs* and classify the *LATs* that their similarity measures are equal to 1 into a subset. Each subset represents a unique object pattern.

To determine if there is a dominant object pattern for performing a specific activity, we calculate the probability of the occurrence of a unique object pattern for all object patterns within the set of *LATs* for the activity. We then specify a threshold value for the probability of occurrence so that when the occurrence probability of a specific object pattern is greater than or equal to the threshold value, the corresponding subset can be viewed as the dominant object pattern. For example, suppose that there are five object patterns for performing an activity and the occurrence probability of the third object pattern is 0.83. This means that the activity is performed 83% of the time using the third object pattern and only 17% using the other patterns. In this case, the third object pattern can be reasonably regarded as the user profile for this specific activity. On the other hand, if all probability values are roughly evenly distributed and each value is very small, it can be assumed that the activity is performed in a random manner and there is not a specific preferred way of performing the activity. In our study, we define 2/3 as the threshold value of the occurrence frequency. Table 4.2 summarizes the variables, their descriptions, and the pseudo code of the algorithm for this object pattern learning method.

Table 4.2 The algorithm for learning object patterns

Variables	Descriptions
SL(z)	*The set of all LATs for the specific activity z*
pop_k	*The probability of occurrence of the object pattern k*
T_{pop}	*The threshold of pop = 2/3*

// discover unique object patterns
set $SL(z)$ and T_{pop} from Phase II outputs
for any LAT_i, $LAT_i \in SL(z)$, **do**
semantic unfolding as illustrated in step 2 in Example 1
enddo
set a counter *uopNum* = 0, which represents the number of the unique object patterns in $SL(z)$
while ($|SL(z)| > 0$)
set LAT_{base} to an arbitrary member of $SL(z)$
create a new subset $SSL(z)_{uopNum}$ with LAT_{base} as the only member
for ($1 \leq i \leq |SL(z)|$)
calculate $sim_{lat}(LAT_{base}, LAT_i)$, where $LAT_i \in SL(z)$
if ($sim_{lat}(LAT_{base}, LAT_i) = 1$)
put LAT_i into the subset $SSL(z)_{uopNum}$
remove LAT_i from $SL(z)$
else
leave LAT_i in $SL(z)$
endif
endfor
Increase the counter $uopNum = uopNum + 1$
endwhile // this will create *uopNum* subsets $SSL(z)$
for ($1 \leq i \leq uopNum$)
calculate $pop_i = |SSL(z)_i| / |SL(z)|$
if $(pop_i) \geq T_{pop}$)
$SSL(z)_i$ represents a dominant object pattern
else
No user profile for this activity
endif
Endfor

4.4.2 Activity Duration Detection

Duration information of an activity model is useful in continuous activity recognition. It helps define the sliding time window for dynamic sensor data segmentation (more details in Chap. 5). It is also a key indicator of a user's behavioural changes, which provide personalised assistance, e.g., specifying the waiting time for a reminder. We calculate duration information using all *LATs* of an activity based on the time points at which the first and last sensor activations of the *LATi* are received. Table 4.3 displays the algorithm for calculating the minimum, maximum, and average duration of a user performing an activity. The algorithm is a continuum of the object pattern learning algorithm in Table 4.2.

4.4.3 Activity Patterns Detection

An activity pattern is crucial for providing proactive personalised activity assistance. For example, if an assistive system knows that a user takes medicine twice a day at 10 am and 5 pm respectively, then it can prompt the user to take medicine at these times. Nevertheless, it is difficult to decide an activity pattern and starting time as most ADLs could be carried out randomly dependent on personal preferences. Even with some kind of regularity, ADLs are most likely performed within a time period rather than at an exact time point. We have developed a two-stage approach to discover an activity pattern and starting time from *LATs*. In the first stage, we calculate the daily frequency of occurrence of an activity, namely the average number of activity occurrences in a day during the period of monitoring. The daily frequency of occurrence is used as a criterion to decide if the activity is carried out on a regular basis. It can be determined based on domain knowledge during the initial *LAT* modelling. For example, it could

Table 4.3 The algorithm for learning activity duration

Variables	Descriptions
t_s, t_e	*the first and last sensor activation times*
Du_{min}, Du_{max}, Du_{ave}	*the minimum, maximum and average duration*
// discover the duration information	
Set Du_{min}=initial value, Du_{max}, *and* $Du = 0$	
for ($1 \leq i \leq \mid SL(z) \mid$) // for all *LATs* of an activity	
if ($Du_{min} > (t_{ei} - t_{si})$) $Du_{min} = (t_{ei} - t_{si})$	
if ($Du_{max} < (t_{ei} - t_{si})$) $Du_{max} = (t_{ei} - t_{si})$	
$Du = Du + (t_{ei} - t_{si})$	
endfor	
$Du_{ave} = Du / \mid SL(z) \mid$	

be 1/7, implying that it covers all weekly activities. A regular activity does not necessarily support an activity pattern. For example, a user makes tea twice a day, every day, however, the activity is always carried out at different times. This is a regular activity but does not have a pattern.

In the second stage, we decide if a regular activity follows an activity pattern. To this end, we first partition the 24 h of a day into a number of fixed-length time slots. For example, if the duration of a time slot is 30 min, then a day can be partitioned into 48 time slots. Second, we map the starting time of all *LATs* of an activity into the corresponding time slots. Third, we calculate the probabilities of the occurrence of the activity within each time slot against the total occurrence of the activity. Based on the probability distribution of occurrence and the threshold values of the occurrence probabilities, we can infer whether or not there is an activity pattern.

Table 4.4 displays the algorithm of learning activity patterns, which is a continuum of the algorithms in Tables 4.2 and 4.3. Three inference rules for learning activity patterns have been defined below, which are explained using the example depicted in Fig. 4.3.

Rule 1: If an activity is a regular activity based on the daily frequency of the activity fo_{lat} *(LAT);* and fo_{lat} *(LAT)* is n $\leq$ 1; and the occurrence probability of the activity in the pth time slot is equal or greater than $Prob_{threshold}$; then the activity has a pattern–it happens once 1/n day(s) in the pth time slot. The starting time S_{time} for the activity pattern can be estimated as the average time of the first sensor activation of all *LATs* within the *pth* time slot. The bath activity in Fig. 4.3 illustrates this case. For example, if fo_{lat} $(bath) = 0.5$, $Prob_{threshold} = 70\%$, as $Prob_{(bath)} = 80\% > 70\%$, then it can be inferred that the bath activity happens once every two days in the time slot starting from 7 pm.

Rule 2: If an activity is a regular activity; and fo_{lat} *(LAT)* is n > 1; and the occurrence probability for each time slot is greater than $Prob_{threshold} \times$ (1/n), i.e., the aggregated occurrence probability in the n time slots is greater than $Prob_{threshold}$; then the activity has a pattern–it is performed n times a day within the n time slots. The starting time S_{time} of the nth occurrences can be estimated as the average time of the first sensor activation of all *LATs* within the nth time slot. The tea activity in Fig. 4.3 illustrated this case, i.e., it happens three times a day in three-time slots with the occurrence probability of each timeslot being greater than 23.3%.

Rule 3: If an activity is a regular activity and the occurrence of an activity is dispersed evenly among a number of time slots k where k is significantly greater than fo_{lat} *(LAT)*; and the occurrence probability in each time slot is significantly less than $Prob_{threshold}$; then the activity is a random activity during a day. As such, it makes no sense to infer the starting time of the activity. The phone call activity in Fig. 4.3 illustrates the nature of a random activity.

Table 4.4 The algorithm for learning activity patterns

Variables	Descriptions
fo_{lat}	*the daily frequency of occurrence of an activity*
Stime	*the starting time(s) of an activity*
prob	*the probability of an activity occurrence in a time slot*
$prob_{threshold}$	*the threshold values for prob*
t_{slot}	*the fixed-length duration of a time slot in minutes*
D	*the duration of activity monitoring in days*

// discover activity patterns and starting time(s)
35. calculate $fo_{lat}(LAT, LAT \in SL(z)) = |SL(z)| / D$
36. partition a day into time slots based on t_{slot}
37. map the t_s of all *LATs* in *SL(z)* into corresponding time slots
38. **for** ($1 \leq i \leq 24 \times 60 / t_{slot}$) // for all time slots
39. $prob_i$ = (number of occurrence in the ith time slot) / $|SL(z)|$
40. **endfor**
41. // apply the pattern learning rules
42. **if** ($fo_{lat} \leq 1$ *and prob* at a time slot $p \geq prob_{threshold}$)
43. *LAT* is a regular activity with an activity pattern
44. *Stime* $= (\sum_{j=1}^{K} t_{ji}) / K$, K = number of occurrence in time slot p
45. **else** ($fo_{lat} \leq 1$ *and prob* at any time slot $p \leq prob_{threshold}$)
46. *LAT is a random activity, no need to calculate Stime*
47. **endif**
48. **if** ($fo_{lat} = n > 1$ and each *prob* at n time slots $\geq prob_{threshold} \times (1/n)$)
49. *LAT* is a regular activity with an activity pattern
50. *Stime*(at the nth occurrence) $= (\sum_{j=1}^{K} t_{ji}) / K$, K = number of occurrence in time slot p_i, $i = 1, 2 \ldots n$.
51. **else** ($fo_{lat} = n > 1$ and all *prob* at n time slots $\leq prob_{threshold} \times (1/n)$)
52. *LAT is a random activity, no need to calculate Stime*
53. **Endif**

4.4.4 Activity Knowledge Model Evolution

Once a new activity is discovered as described in Sect. 4.4.3, it is necessary to decide the location of the activity in the hierarchy of the activity ontologies and also an appropriate label that should be assigned to the activity. The label should be meaningful and compliant with other activities' labelling rationale and also the ontological modelling conventions so that it can be easily referred to and understood later. The location of a newly discovered activity in the ontological activity hierarchy can be recommended through the subsumption reasoning of the *UAT* description.

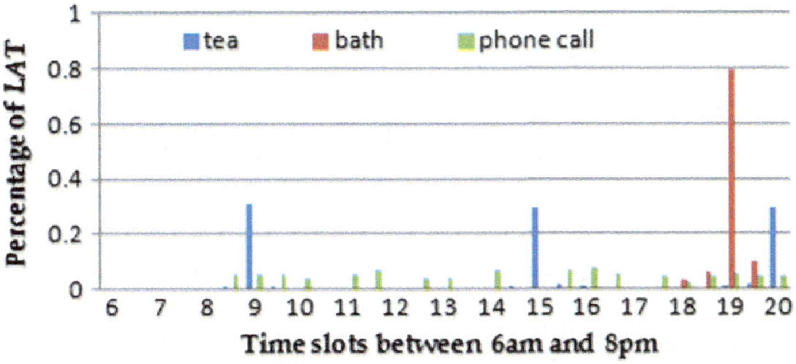

Fig. 4.3 Probability distribution of three ADLs over a period of time

Nevertheless, human intervention is required to validate and finalise the position and label of an activity model in order to maintain the quality of the model. As such, the classification and naming process have been carried out manually using the standard practice of ontological engineering, i.e., a knowledge engineer encodes the new activities and edits the ontologies using an ontology editor.

Similarly, once a user's behavioural features, i.e., activity profiles, are learnt as described in the previous subsections, the activity models should be evolved to reflect the unique manner a user performs activities, e.g., for the purpose of personalised assistance. Given that a user's activity profile is equivalent to an instance of a generic activity model, i.e., an ontological activity class, and for any *LAT* there is a corresponding ontological activity class, activity profile evolution amounts to creating a new instance or updating an existing instance. This can be undertaken automatically by using the standard APIs of the underlying semantic frameworks.

4.5 An Example Case Study

When an actor interacts with objects in sequence in real time, sensor activations are continuously fed into the system. To test and evaluate the presented approach we have created the seed activity ontologies in Phase I using the Protégé ontology editor – refer to Fig. 4.2, through knowledge engineering practice. We have implemented a feature-rich system for activity recognition and model learning in Phase II, as shown in Fig. 4.4. Full detail of the system implementation can be seen in Chap. 9 and Sect. 9.2. When an actor interacts with objects in sequence in real time, sensor activations are continuously fed into the system. Sensor data series are dynamically segmented, will be discussed in next two chapters, and recognition operations are repeatedly performed to carry out continuous, progressive activity recognition. As depicted in Fig. 4.4, the system can dynamically display the activated sensor sequence, the incrementally recognised activities and the system status in real time.

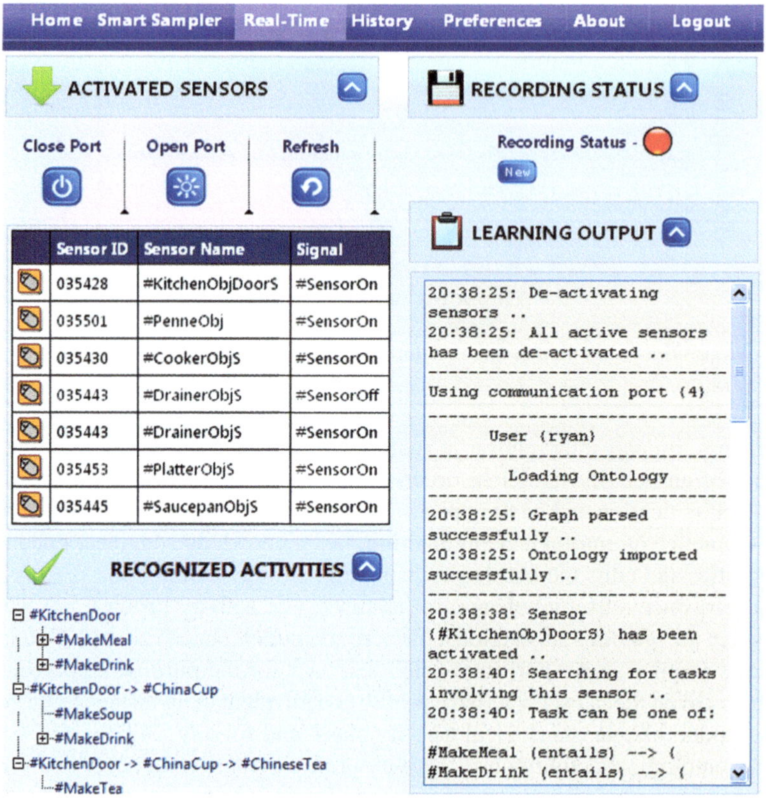

Fig. 4.4 The system interface operating in real time mode

Figure 4.4 depicts the system interface operating in the real time mode. In the left-hand side, the top panel is used for communication port setup; the middle panel displays the sequence of activated objects; and the bottom panel presents progressively recognised activities in a tree-like hierarchy. In the right-hand side, the top panel contains function buttons for data recording and playback; the bottom panel presents a temporal trace of events during the system operation. The system can import activity ontologies, specify reasoning and learning parameters, select the modality of audio reminder, configure hardware and define event priorities and user activity profiles (Table 4.5).

4.5.1 Experiment Design and Data Collection

To systematically test and evaluate activity and profile learning in Phase III, eight typical ADLs as presented in Table 4.6, were selected for the purposes of experi-

Table 4.5 Two examples of activity specifications

Activity		Activity specification (sequence of user-object interactions)
Maketea	TP1	GetCup, GetTea, PourWater, GetMilk, GetSugar
	TP2	GetCup, **PourWater**, **GetMilk**, **GetTea**, GetSugar
	TP3	**GetCup***, GetTea, PourWater, GetMilk, GetSugar
BrushTeeth	TP1	RunSink, GetToothbrush, GetToothpaste, GetMouthwash
	TP2	**GetToothbrush**, **GetToothpaste**, **RunSink**, GetMouthwash
	TP3	RunSink, GetToothbrush, **getSoap****, GetToothpaste, GetMouthwash

*faulty sensors that do not fire; ** false or extra sensor reading

mentation. For each activity, the required objects for performing the activity were identified and for each of them a contact sensor was attached. Each activity was designed to be performed in three different ways, leading to three different types of activity specification as illustrated in Table 4.5. The Type 1 activity specification, namely TP1 in short, can be viewed as the "standard" way of performing a specific activity. The Type 2 activity specification has the same set of objects; however, they are interacted with in a different order. The Type 3 activity specification has a different set of objects as it is intended to simulate noise on the sensor data, i.e. a faulty sensor by omitting a user-object interaction or a false sensor reading by adding an irrelevant object interaction. In addition, in order to test the activity learning capability, we deliberately remove activity models, *MakeChocolate* and *BrushTeeth*, two of the eight selected activities from the seed activity ontologies.

Three actors took part in the experiments. Each of the participants interacts with the objects of each activity of the eight activities in accordance to the activity specifications for two rounds. This leads to a total of 3 (types) $\times$ 8 (activities) $\times$ 2 (rounds) $\times$ 3 (actors) $=$ 144 action traces. After activity recognition in Phase II the system produced 100 *LATs* and 44 *UATs* as presented in Table 4.6.

4.5.2 Analysis and Evaluation

Our evaluation has focused on the performance of learning distinct activities from *UATs*, and the performance of discovering the dominant object pattern from *LATs* in activity profile learning. This is because semantic based similarity calculation and classification are the central underpinning mechanisms for the presented methods. In addition, evaluation of time-related metrics, e.g. duration or activity patterns will only make sense if the data are generated by real users performing real ADLs over a relatively long period of time. This has been proven to be difficult due to technical, privacy and ethical issues. Furthermore, temporal information in these learning

Table 4.6 Recognition results of the 144 activities

Activities		Actor1		Actor2		Actor3		Sum L/U
Exp		1	2	1	2	1	2	
Make tea	TP1	L	L	L	L	L	L	6/0
	TP2	L	L	L	L	U	L	5/1
	TP3	L	L	L	U	U	U	3/3
Brush teeth	TP1	U	U	U	U	U	U	0/6
	TP2	U	U	U	U	U	U	0/6
	TP3	U	U	U	U	U	U	0/6
Make coffee	TP1	L	L	L	L	L	L	6/0
	TP2	L	L	U	L	U	L	4/2
	TP3	L	L	L	L	L	L	6/0
Have bath	TP1	L	L	L	L	L	L	6/0
	TP2	L	L	L	L	L	L	6/0
	TP3	L	L	L	L	L	L	6/0
Watch TV	TP1	L	L	L	L	L	L	6/0
	TP2	L	L	L	L	L	L	6/0
	TP3	L	L	L	L	L	L	6/0
Make chocolate	TP1	U	U	U	U	U	U	0/6
	TP2	U	U	U	U	U	U	0/6
	TP3	U	U	U	U	U	U	0/6
Make pasta	TP1	L	L	L	L	L	L	6/0
	TP2	L	L	L	L	U	L	5/1
	TP3	L	L	U	L	L	L	5/1
Wash hands	TP1	L	L	L	L	L	L	6/0
	TP2	L	L	L	L	L	L	6/0
	TP3	L	L	L	L	L	L	6/0
Sum L/U	All TPs	18/6	18/6	16/8	17/7	14/1010	17/7	100/44

Here TP—the type of activity, Exp1 and Exp2—the two rounds of experiments respectively, L and U—a LAT and UAT respectively, and Sum—the number of L and U for a particular type of activity and a particular actor respectively

Table 4.7 The activity discovery results from UATs

	Ground truth		UAT Subsets SSU_i
Activity label	UAT	LAT	Total 21 subsets, SSU_1–SSU_{21}
MakeChocolate	18	0	12 in SSU_1, 1 in each $SSU_{6–11}$
MakeTea	4	14	1 in each $SSU_{18–21}$
MakeCoffee	2	16	2 in SSU_3
BrushTeeth	18	0	12 in SSU_2, 1 in each $SSU_{12–17}$
MakePasta	2	16	1 in each $SSU_{4–5}$

Please refer to Table 4.5 for the definition of SSU_i

methods is mainly used for numerical calculation, i.e. the duration, starting time and frequencies, which has already clearly illustrated in previous discussions.

Results and analysis on learning new activities: We apply the activity learning algorithm in Table 4.4 to the UAT dataset in Table 4.6 to learn new activities. Table 4.7 displays the activity learning results. The "Ground Truth" column presents what actually happened in the experiment whereas the "*UAT Subset*" column lists the classified subsets of the 44 UATs. Among six modelled activities three of them, i.e. *WashHands*, *WatchTV* and *HaveBath*, have been fully recognised without generating any UATs, so they are not listed in Table 4.7. The other three modelled activities, i.e. *MakeTea*, *MakePasta* and *MakeCoffee*, have generated four, two and two *UATs* respectively. This is because we randomly introduce a sensor noise into the Type 3 activity specification, the activity traces from TP3 may be recognised or not depending on the nature of the noise, thus leading to *UATs*.

For the two unmodelled activities, *MakeChocolate* and *BrushTeeth*, each consists of 18 *UATs* which are classified into 7 subsets *SSU*. One subset has 12 *UATs* and other six subsets each have one *UAT*. This is because both Type 1 and Type 2 activity specifications use the same set of objects, thus leading to 12 *UATs* in *SSU1*. The Type 3 activity specification simulates random sensor noise by introducing an irrelevant object into the activity, thus leading to 6 different action traces. The comparison between the *UAT* classification results and the ground truth proved that the semantic similarity based *UAT* classification is 100% accurate in terms of similarity criteria $sim_{uat}(UAT_i, UAT_j) = 1$. In the case that the duration of observation *D* is available, it is straightforward to follow the activity learning algorithm to identify the distinct regular activities.

We apply the algorithm in Table 4.2 to all LATs in Table 4.6 to learn object patterns. Table 4.8 presents the analysis results for three of the six modelled activities. From left to right the first and second columns contain the activities and the total number of *LATs* in the corresponding activity. The third column displays the unique object patterns among all *LATs* of the activity while the fourth one shows the number of *LATs* for each unique object pattern. The fifth column presents the probabilities of occurrence of a unique object pattern. As can be viewed from the results, each activity has two major activity patterns with a similar percentage of occurrences. In

Table 4.8 Part of the activity learning results from LATs

Activities	*LAT* No.	Unique object oatterns (*UOP*)	*LAT* No. for each *UOP*	pop_x *(%)* for each *UOP*
MakeTea	14	UOP_1	6	42.86
		UOP_2	5	35.71
		UOP_3–UOP_5	1	7.14 each
MakePasta	16	UOP_1	6	37.5
		UOP_2	5	31.25
		UOP_3–UOP_7	1	6.25 each
WashHands	18	UOP_1	6	33.33
		UOP_2	6	33.33
		UOP_3–UOP_9	1	5.55 each

addition, a number of patterns are also identified for each activity with each pattern having only one *LAT*. The learning results are in line with the ground truth of the experiment. The two major activity patterns correspond to the Type 1 and Type 2 activity specifications. The occurrence of a number of one-*LAT* pattern in each activity corresponds to the Type 3 activity that is performed randomly by introducing random noise, thus no sequence of objects are identical. The matching of the analysis results with the ground truth of the experiment proves the method for learning object patterns is effective.

There are a number of object patterns for each activity in Table 4.8. This is because the activity specifications are deliberately designed to contain two major object patterns, i.e. Type 1 and Type 2, and a number of random patterns in Type 3, to test and evaluate various aspects of activity and profile learning methods. In a real situation a user may have one dominant object pattern or simply perform in a random way. Nevertheless, the experiments and analysis results demonstrate the learning method and process. For example, if we set the threshold of the probability of occurrence of the object pattern (pop_x) to 36%, then the unique object pattern UOP_1 for both *MakeTea* and *MakePasta* will be identified as the dominant object patterns. For the *WashHands* activity there is no object pattern.

Sensor noise such as faulty sensors, communication and processing errors is inevitable in real use scenarios. In our experiments we simulate sensor noise in Type 3 activity specifications, leading to six occurrences of sensor noise for each activity among its eighteen activity occurrences, equivalent to 33.33% data accuracy. As can be seen from the results in Table 4.8 a sensor noise does not have to affect activity recognition, i.e. generating an UAT. It will be up to the nature of sensor noise that determines whether or not an action trace with a sensor noise could be recognised. The impact of sensor noise on recognition accuracy has been discussed in [13].

Sensor noise: affects activity and profile learning. The analysis results in Table 4.7 show that the two unmodelled activities, *MakeChocolate* and *BrushTeeth* each have 18 UATs but only 12 of them are classified into one set due to sensor noise, equivalent to a 66.67% classification rate, which is resulted from our simulation of sensor noise for exactly one third of activities in the experiments. Nevertheless, the extent to which the noise affects the classification rate is dependent on the similarity threshold which is used to decide whether or not two traces are deemed as similar. For example, our study only classifies absolutely similar traces, i.e. $sim_{uat}(UAT_i, UAT_j) = 1$, into a set. If we reduce the similarity threshold, e.g. to 0.8, then any traces with $simuat(UAT_i, UAT_j) \geq 0.8$ will be classified to the same set. In this case the classification rate (66.67%) and the noise level (33.33%) will both be changed. This actually means that two activity traces with one of them having sensor noises such as a missing sensor observation or a wrong object can still be classified to a set if other objects are the same. Understandably, the lower of the similarity threshold, e.g. 0.65, the more sensor noises can be accommodated. From this perspective, activity learning is not sensitive to sensor noises supposed the noise level is relatively low.

Though the determination of appropriate similarity threshold which is essentially to decide the noise level the presented approach can handle requires further investigation, current study has shown that our approach itself is conceptually and theoretically correct without specific limitations. The impact of sensor noise on profile learning as depicted in Table 4.8 can be discussed and explained in the same way as above. We shall not elaborate here due to limited space.

Computational performance: In the 3-phase iterative process of the hybrid approach to activity model learning, real-time continuous activity recognition requires high computational performance to ensure dynamic on-the-fly situation generation and reasoning against the activity models. The experiments and evaluation in Chap. 3 have shown the computational performance for real-time activity recognition is satisfactory. As activity and profile learning are intended to be performed periodically offline, and most computation in these learning algorithms involves linear time complexity with regard to dataset volume, we believe that the technical correctness of these learning algorithms is more important than their computational performance. As such, our experiments and evaluation have focused on technical assessment.

Knowledge-driven versus data-driven: The presented hybrid approach combines knowledge-driven manual model specification with data-driven automatic model learning. One question arising from the study is to what extent models should be manually specified in advance. Should we specify as many models as possible with few to be learnt or the reverse? Relying on manual specification too much will have the disadvantages of the knowledge-driven approach. On the other hand, relying on automatic model learning too much will have the drawbacks of data-driven approaches. While the approach allows flexible specification of the initial seed activity models, it is an interesting research question on how to strike an optimal balance between the two ways of activity modelling.

Experiences and initial findings from our current studies suggest that we should specify as many generic coarse-grained activity models as possible as the models at this level of abstraction are generic and applicable to all users, thus insensitive to low-level special behaviour of individual users. On the other hand, we should learn as many fine-grained activity models as possible as the models at this level of abstraction reflect the uniqueness and dynamics of an individual user's behaviour.

Data-driven activity learning plays more important role in improving activity model accuracy and addressing the changing nature of activity models.

4.6 Summary

This chapter introduced a hybrid approach to creating complete, accurate activity models through incremental activity discovery and profile learning. We have described a 4-phase iterative process and discussed the methodology of each phase of the lifecycle. While Chap. 3 described the details of ontological activity modelling and recognition, this chapter has presented the details of activity and profile learning methods by which activity models can be expanded, personalised and adapted. The compelling feature of the approach is that it combines the strengths of traditional data mining-based activity modelling with that of ontology based explicit activity modelling, making our approach flexible, applicable and scalable in terms of reusability, rapid system development and deployment. We implemented the approach in a feature-rich assistive system and conducted systematic controlled experiments in a number of well-designed activity scenarios. Initial results have demonstrated that the approach and algorithms are technically correct, viable and robust. Though the experiment dataset is not very large, it is representative and serves the purposes well.

References

1. WHO | International Classification of Functioning, Disability and Health (ICF). WHO (2018)
2. Liao L, Fox D, Kautz H (2007) Hierarchical conditional random fields for GPS-based activity recognition. In: Thrun S, Brooks R, Durrant-Whyte H (eds) Robotics research. Springer, Berlin, pp 487–506
3. Lester J, Choudhury T, Kern N, Borriello G, Hannaford B (2005) A hybrid discriminative/generative approach for modeling human activities. In: Proceedings of the 19th international joint conference on artificial intelligence. Morgan Kaufmann Publishers Inc., San Francisco, CA, USA, pp 766–772
4. Hu DH, Yang Q (2011) Transfer learning for activity recognition via sensor mapping. In: IJCAI international joint conference on artificial intelligence
5. Rashidi P, Cook DJ (2011) Activity knowledge transfer in smart environments. Pervasive Mob Comput
6. Van Kasteren TLM, Englebienne G, Kröse BJA (2010) Transferring knowledge of activity recognition across sensor networks. In: Lecture notes in computer science (including subseries Lecture notes in artificial intelligence and lecture notes in bioinformatics)
7. Cook D, Feuz KD, Krishnan NC (2013) Transfer learning for activity recognition: a survey. Knowl Inf Syst 36:537–556
8. Perkowitz M, Philipose M, Fishkin K, Patterson DJ (2004) Mining models of human activities from the web. In: Proceedings of the 13th conference on World Wide Web - WWW '04
9. Tapia EM, Choudhury T, Philipose M (2006) Building reliable activity models using hierarchical shrinkage and mined ontology. In: Lecture notes in computer science (including subseries Lecture notes in artificial intelligence and lecture notes in bioinformatics)

10. Mann CJH (2003) The description logic handbook – theory, implementation and applications. Kybernetes (2003)
11. Jain AK, Dubes RC (1988) Algorithms for clustering data, Englewood Cliffs, N.J.: Prentice Hall, ISBN:0-13-022278-X
12. Witten IH, Frank E, Hall MA (2011) Data mining: practical machine learning tools and techniques, 3rd ed., Elsevier, ISBN:978-0-12-374856-0
13. Hong X, Nugent CD (2013) Segmenting sensor data for activity monitoring in smart environments. Pers Ubiquitous Comput 17:545–559

Chapter 5
Time-Window Based Data Segmentation

5.1 Introduction

Dense sensor-based activity monitoring and recognition has received increasing attention in SH environments due to privacy and ethical considerations. In such environments, sensors are attached to objects in the environment, e.g. fridges, cupboards, and an inhabitant's interactions with these objects are monitored and used to identify the ongoing activities of daily living (ADLs). A key problem in dense sensor-based activity recognition when sensors are activated along a timeline is how sensor data are segmented so that the set of sensor interactions represents exactly a unique activity.

Existing works on knowledge-driven activity recognition do not clearly articulate the mechanism about how and what sensor data are selected from a live data stream for performing activity inference. In some research experiments that support online continuous activity recognition, the experiments restart manually each time an ADL is identified. For the approach to be applicable to real-world use scenarios it is necessary that after an ADL is identified, the activity recognition process should continue on fresh sensor data and decide what to exclude from those already used in the previously identified ADL(s). Obviously, this is not a trivial task and requires the development of a suitable discriminating strategy.

To dynamically decide an appropriate set of sensor data from a live sensor data stream for real-time activity recognition, we introduce a segmentation approach that makes use of temporal information associated with sensor data and temporal characteristics of an activity. The approach addresses two important issues: 'segmentation and aggregation' and 'the conditions that trigger ontological reasoning'. Segmentation breaks down a sensor data stream into fragments that can be mapped to activity descriptions, while aggregation combines a finite collection of sensor data items available in a segment for activity inference. As such, the approach is able to support continuous segmentation and aggregation along a timeline, thus allowing real-time ongoing activity recognition. To achieve this goal, in this chapter we (1) describe a time window based segmentation model and related algorithms for real-time onto-

L. Chen and C. D. Nugent, *Human Activity Recognition and Behaviour Analysis*, https://doi.org/10.1007/978-3-030-19408-6_5

logical activity recognition; (2) develop various mechanisms for dynamically manipulating time window parameters; (3) implement the proposed model and reasoning algorithms; (4) develop tools to obtain temporally-rich ADL data for testing and evaluation; and (5) evaluate the performance of the proposed model and algorithms in supporting real-time activity recognition. The research is based on typical ADL activities that an inhabitant can perform in the kitchen, lounge and bathroom of a smart home, e.g. cooking, watching television, and showering.

This chapter presents a development of a systematic approach to dynamic sensor data segmentation for real-time continuous activity recognition. This dynamic sensor segmentation approach number of benefits. Firstly, a time window-based segmentation model that is applicable to a wide range of activity recognition scenarios. Secondly, we illustrate various mechanisms for dynamic model parameter manipulation during activity recognition, such as the setting, shrinking, and expansion of the time window's length, thus adapting the segmentation model in terms of the way activities are performed. Thirdly, we integrate the dynamic sensor data segmentation approach into an ontology-based algorithm for real-time, progressive activity recognition. This provides a basis for the implementation of re-usable knowledge-driven algorithms and applications for real-time activity recognition. In addition, we develop a synthetic ADL data generator that can be used to quickly generate temporally-rich synthetic ADL data for evaluation of activity recognition algorithms. It is believed that the time window based segmentation model and associated algorithms in activity recognition provide a realistically scalable, reusable approach to continuously recognising activities of different complexities in a Smart Home context.

5.2 Recent Work on Temporal Data Segmentation

The issue of sensor data segmentation in knowledge-driven activity recognition has received little attention in existing work. For instance, in [1] the authors present a knowledge-driven activity recognition approach but do not provide the details of the method for sensor data selection. Despite showing that ontology-based activity recognition is feasible, the absence of a suitable method for sensor data selection makes the presented method difficult to replicate. However, another knowledge-driven activity recognition method presented in [2] uses competing hidden Markov models to segment a sensor data stream. The selected sensor data is used to perform spatiotemporal and context reasoning for activity recognition. They use a variable window length and the window moves over a sequence of observations. The main weakness of this approach is that it requires a pre-existing dataset to determine the optimal size of the time windows and the segmentation rules that it uses. Since the same individual or different individuals may perform the same activity in many different ways, this method will be difficult to reuse. In addition, the derived optimal window lengths have to be revised to deal with new situations.

The use of one-minute time slices to evaluate the effectiveness of ontology-based activity recognition is presented in [2]. Sets of sensor data are selected every minute for activity inference. The work is based on van Kasteren dataset [3] and the main limitation is that the ontology used is modelled on and closely tied to the dataset making it difficult to re-use. In addition, the fixed-size time slices used may lead to a huge computational expense since the activity inference engine is forced to periodically sample the data stream even when no new sensors have been activated. Furthermore, its ability to support real-time activity recognition has not been discussed. Ontologies and video are used for activity recognition in [9], whereby an ontology-based knowledge base supports the recognition of human activities from video sequences. The ontology models human activities, in terms of entities, environments and interactions, and creates semantic links between events and activities. Vision-based techniques are used to select the input data for activity inference based on a pre-existing dataset. Our work is modelled on a dense sensing framework, and as a result the computer vision based techniques used in [4] are less suitable. However, the authors adopt a method to select the input data used in activity recognition which is comparable to the problem that we aim to address, i.e., to select a subset of sensor data for activity inference. From the foregoing, it is clear that in most knowledge-driven activity recognition work the method used to select sensor data is either non-existent or, at best, ad hoc. There is a need to develop a systematic approach that can be applicable in different knowledge-driven activity recognition approaches to help segment and then aggregate sensor data.

In the data-driven activity recognition community, the problem of sensor data segmentation has been widely explored has discussed in Chap. 2 (Sect. 2.2). The notion of time windows is adopted to provide a basis for handling time-dependent data, e.g., the sensor data stream. However, some sensor data segmentation approaches use static sliding windows to segment the data stream [3, 5, 6] while others use dynamically derived time window lengths [7–9]. The notion of time slices is used in [10] to derive segments used to perform activity recognition. In [5, 6], a sliding window method is used to derive features used in activity inference by the proposed algorithms. The time windows used in [5] are made to have 50% overlap. The main criticism for static sliding windows is that incorrect lengths can truncate an activity instance or overlap activity instances leading to recognition failure. Due to the above problem, this chapter is closely related to [7–9] whereby time window parameters are varied.

The work in [7] uses temporal information (i.e., the average activity duration) to set different length values to the time windows at initialization; however, once a time window is activated its length cannot be dynamically modified. This can cause the time window to overlap the end of one activity and the beginning of the next one, thus leading to recognition failures. The notion of time contiguity in sensor data is used together with location context to segment sensor data from state change sensors in [8]. Any noted changes in location context between two consecutive sensors are used to signify a break point. The break point then helps identify the start and end of segments on which activity inference is eventually performed by evidential fusion [11]. This approach will work well when consecutive activities occur in different locations; however, segmenting the sensor data stream arising from the same location may prove difficult if the break point is not detected. Since, recognition is only attempted on a segment after the start and end points are identified, this approach assumes that the user always performs activities correctly, making diagnosis that is necessary for activity assistance difficult or impossible.

Although our work is in the area of knowledge driven activity recognition, it is slightly similar to the work in [9] since it uses dynamic windows. In [9], the length of the window is dynamically derived at runtime based on the occurrence of specified low-level events, e.g. the change of sensor state. The key difference with our work is that while they use primitive events to dynamically manipulate time window parameters, we use high level context information such as activity duration, and the current status of recognition resulting from a high-level activity inference event. As a result, our approach is able to utilize high quality knowledge in sensor data segmentation.

Following the above discussion, we contend that by capturing some temporal features of sensor data and by extension that of activities, the sensor data stream can be broken into segments for real-time activity recognition. We use dynamically varied time windows based on the temporal information of activities to support this segmentation, and activity recognition is then performed on these segments. The main strengths of the proposed approach are that it is systematic, simple, well-defined and easy to implement; and applicable to any user or dataset; hence, can be replicated.

5.3 Real-Time Activity Recognition Analysis

5.3.1 Concept and Architecture

Continuous, real-time activity recognition helps to identify ongoing ADLs as they occur, thus offering the possibility to provide timely assistance for SH inhabitants. In ontology-based activity recognition, when an ADL is performed along a timeline, the contextual information associated with the ADL is captured incrementally and subsumption reasoning is used to infer the ongoing ADL. At the initial stages of an ADL, subsumption reasoning may only classify the contextual information to a generic ADL class. However, as more contextual information is obtained over time, and reasoning is continuously performed, it would be possible to recognize the specific ongoing ADL.

In a dense sensing based SH, contextual information is captured through a variety of sensors, with each sensor representing a particular view of the prevailing situation. From the activated sensors, an agent can infer physical and contextual entities, e.g. objects, locations, times, and events. For example, a pressure sensor can be attached to the sofa in the lounge. Given that this knowledge is explicitly encoded in the ADL ontology, whenever this pressure sensor is activated it is possible to infer that the inhabitant is in the lounge and is sitting on the sofa. Consequently, this allows the inference of an activity that occurs in the lounge while sitting on the sofa, e.g. reading a book or watching television. Since most ADLs require the fusion of data from multiple sensors over time to infer high-level activities, it is necessary to first aggregate a sequence of sensor activations in order to generate a situation at a particular time point during activity recognition. For a more detailed description of the ontology-based activity recognition approach, we refer the interested reader to [4] due to space limitations.

While work in [12] described the rationale and algorithm for semantic reasoning for activity recognition, it did not present any details about sensor data segmentation that is critical for continuous real-time activity recognition. In this chapter, we focus on extending the ontology-based approach in [12] with a sensor data segmentation mechanism so as to support real-time activity recognition. Figure 5.1 shows the three-layer architecture for the extended approach, namely context selection, iterative action inference, and activity recognition layers.

Whenever activities are performed, the incoming sensor data is received as a sensor data stream and segmented in the context selection layer. Context selection refers to the process by which the stream is divided into a set of fragments using temporal segmentation. In temporal segmentation, the stream is analysed along a temporal dimension using the temporal properties (of both sensor data and activities) captured

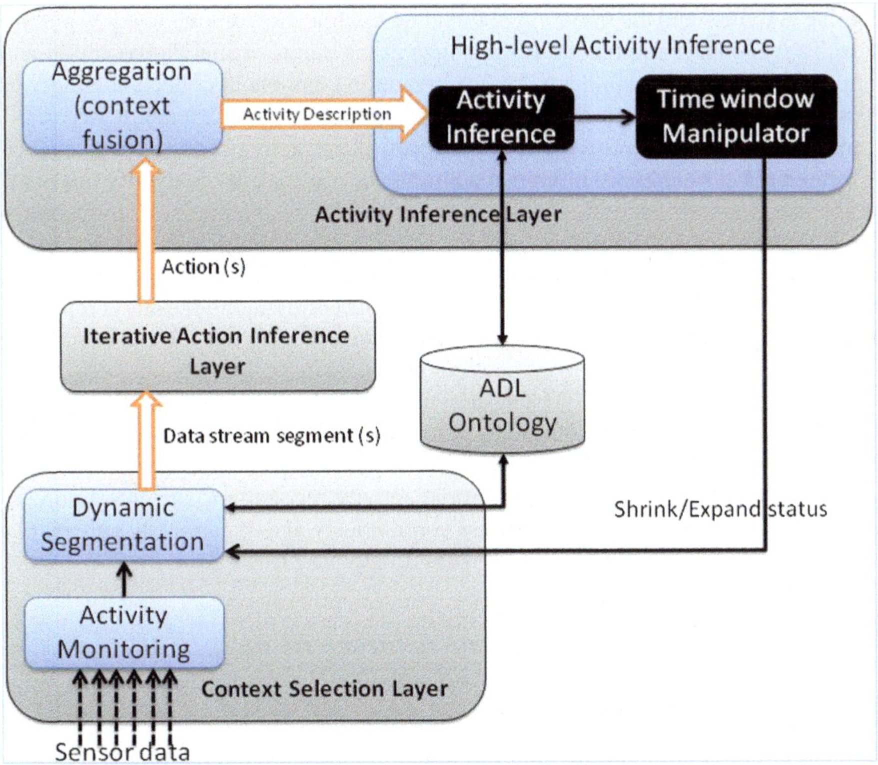

Fig. 5.1 Architecture of real-time activity recognition approach

by time windows. This ensures that only those sensor activations occurring within a given time window are included in the segment. To achieve its goal, this layer uses activity monitoring and dynamic segmentation components. The activity monitoring component allows sensor data to be received, while dynamic segmentation component implements the temporal segmentation algorithm. We model the notion of a time window as a data structure made up of a number of parameters. Section 5.3 provides the formal model of the time window mechanism, provides a detailed description, and the algorithm that utilizes it in activity recognition.

The iterative action inference layer analyses the segments of the data stream that are generated in the context selection layer to identify a collection of actions (also called low-level activities) associated with the activated sensors. Typically, a time window's sensor activations are processed against the ADL ontology to determine the ongoing primitive action, e.g. '*cup is used*'. This action can be represented as context information in the ontology by a property assertion that is equivalent to the description: '*hasContainer* property associated with unknown ADL activity *X* has value *cup*'. This process is repeated for all sensor activations that have so far been received in the time window. A collection of such low-level (simple) activities may combine to constitute the activity description of one or more high-level (complex) activities. The activity inference layer is responsible both for the inference of ongoing activities and the initiation of dynamic modification of time window parameters. To this end, it is made up of two main components, namely aggregation and high-level inference components. The aggregation component collects the individual property assertions from the iterative action inference layer together to derive the overall description of the current activity. The resulting activity description is passed on to the high-level activity inference component. The high-level activity inference component is made of activity inference and time window manipulator components. The action inference component uses the activity description, ADL ontology, and ontological reasoning to infer the ongoing ADL, e.g. '*make tea*'. If a specific ADL is inferred, the recognition process is considered successful and the result is reported. Otherwise, a generic ADL is reported, and the system will wait for additional sensors to be activated before attempting recognition again. In this way, ongoing ADLs can be progressively inferred. The system can dynamically initiate the shrinking or expansion of time windows whenever necessary through the time window manipulator component. The mechanism for shrinking and expansion is described in the next section. To ensure perpetual, real-time activity recognition, the entire process continues to run with new time windows continuously and dynamically generated.

5.3.2 Data Stream Segmentation Characterisation

A key factor in continuous real-time activity recognition is how to select the set of activated sensor data to be aggregated for activity classification. In a typical Smart Home, sensors will be continuously activated, and the resulting sensor data sequence needs to be broken down into fragments that can be mapped to specific ADL activities. To segment a sensor data stream, this work presents a number of scenarios and configurations that can be considered. The scenarios are divided into two main categories: overlapping and non-overlapping time windows. In overlapping time windows, two or more distinct time windows can share some activated sensors. On the other hand, whenever non-overlapping time windows are used, no single activated sensor is shared by two or more time windows.

Under each category, there are four scenarios to be considered. The first scenario uses fixed-sized time windows, whereby all time windows are created of the same size (i.e., given the initial time window has size w_0, any newly created window will also have its size set to w_0). The second scenario uses variable-sized time windows. In this case, the lengths of newly created time windows are dynamically derived at run-time such that the length of any new window is a multiple of that of the initial window (i.e., given the initial window has length w_0, any new window will have its length set to $a*w_0$, where a is a positive real number). Regarding both scenarios, a key challenge is how to choose optimal sizes at runtime. The third and fourth scenarios allow time windows to be dynamically shrunk and/or expanded at runtime as a result of activity inference. Scenario three is a variation of scenario one because it uses fixed-sized time windows. Similarly, scenario four is a variant of scenario two. A key challenge is the criteria for triggering shrinking or expansion of a time window. The resulting eight distinct configurations are depicted in Fig. 5.2a–h.

From the scenarios provided, it is clear that although the task of segmenting a sensor data stream with time windows is complex, there are various methods that can be used to achieve it. However, choosing the most suitable method for segmentation is a non-trivial task. Providing support for the different configurations described requires careful design of the time windows together with an appropriate choice of the parameters and strategies for manipulation. In the next sections, we present a time window-based approach and algorithms that model and implement the presented scenarios.

5.4 Sensor Data Segmentation Modelling

This chapter presents a mechanism that uses time windows to decide on which sensor activations to use for activity inference. The mechanism utilizes a sensor data segmentation model that is modelled by a time window data structure. To implement the different configurations shown in Fig. 5.2, the time window data structure provides various parameters. Some parameters can be present but can remain unchanged or be varied, while others are dynamically set at runtime. The time window model and its parameters are described in the next section.

To illustrate the use of the time window model, the scenario in Fig. 5.2e that allows windows to be shrunk and/or expanded was selected. The lines marked in the form $TW - N'$ indicate that the corresponding initial time window has been shrunk or expanded. In the diagram, we have two windows that are not modified (TW-1, TW-4), two that are shrunk (TW-0, TW-2), and one window (TW-3) that is

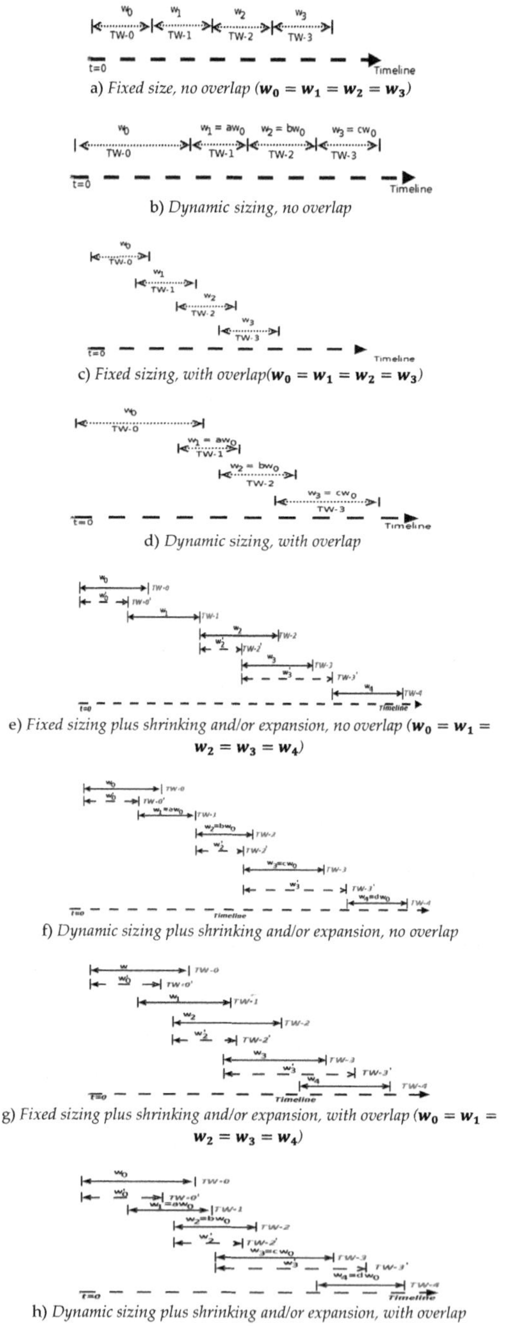

a) *Fixed size, no overlap* ($w_0 = w_1 = w_2 = w_3$)

b) *Dynamic sizing, no overlap*

c) *Fixed sizing, with overlap*($w_0 = w_1 = w_2 = w_3$)

d) *Dynamic sizing, with overlap*

e) *Fixed sizing plus shrinking and/or expansion, no overlap* ($w_0 = w_1 = w_2 = w_3 = w_4$)

f) *Dynamic sizing plus shrinking and/or expansion, no overlap*

g) *Fixed sizing plus shrinking and/or expansion, with overlap* ($w_0 = w_1 = w_2 = w_3 = w_4$)

h) *Dynamic sizing plus shrinking and/or expansion, with overlap*

Fig. 5.2 Representation of time-based sensor data segmentation scenarios expanded. In the current work, only non-overlapping time windows are investigated. During expansion, the time window's length is extended so as to accept further sensor data. This occurs whenever additional sensor data is required to successfully infer an activity but the pending window length would be inadequate to cover the given activity's duration as provided in the activity ontologies. In addition, a window can be expanded whenever a generic activity has been identified but the pending window length is inadequate and thus requiring additional time to successfully infer any of the specific descendants of the activity.

Conversely, during shrinking the time window is truncated before its present length is exhausted. Typically, as soon as an ADL is recognized, the ADL ontology is used to determine whether additional sensor activations should be anticipated or not. In addition, the identified activity's duration information available in the activity ontologies can also be used to determine if the time window should be truncated. If there are no further sensor activations expected or the duration has been exhausted, the current time window is closed and a fresh time window activated. Alternatively, if it requires additional sensor activations or the duration has not been exhausted, then the time window will continue to be active until either its present length is exhausted or further sensor activations are obtained. In addition, an assistive system can be invoked to provide some interventions, e.g. prompts and suggestions, to the user in an Ambient Assisted Living (AAL) environment.

5.4.1 *Formal Time Window Modelling*

A formal time window model is proposed whose characteristics and operation are described below. We define a number of parameters to describe how the time window is manipulated. Significant parameters include start time, end time, window length, the enclosed collection of sensor data, and overlap, shrinking and expansion capabilities. Other parameters are used to provide a means for manipulating the time window data structure. Some (dependent) parameters (e.g. end time) are assigned dynamically during recognition processes, while the independent ones (e.g. window length) are present.

Let:

- α: the start time for a time window.
- α: the end time for a time window.
- w: the length of a time window.
- Ω_α: a time window whose start time is α.
- $\overline{\overline{\Omega}}_\alpha$: sensor data set. This is a data structure for storing the set of sensor data belonging to a given time window.
- $\vec{A}$: a vector of activity labels assigned to the time window after activity inference.
- γ: reasoning start mode. Used to determine when to trigger activity inference.
- ρ: time window factor. Used to derive the size of a new time window from the initial time window.

- μ: sliding factor. Used to determine the size of the slide applied to the active time window to move it over the sensor data stream.
- δ: change factor. Used to determine the magnitude used to expand or shrink the length of a time window.

We can define a time window, Ω_α, as a 9-tuple with nine properties: $\overline{\overline{\Omega}}_\alpha$, α, ω, w,γ, ρ, μ, δ, and $\vec{A}$ as shown in the expression below:

$$\Omega_\alpha :< \overline{\overline{\Omega}}_\alpha, \alpha, \omega, w, \gamma, \rho, \mu, \delta, \vec{A} > \quad (5.1)$$

The end time, ω, can be computed from the start time, α, and window length, w, as shown below:

$$\omega = \alpha + w \quad (5.2)$$

Given that a sensor activation arriving at time, t, is denoted by sa_t; $\overline{\overline{\Omega}}_\alpha$ can be defined below:

$$\overline{\overline{\Omega}}_\alpha : \left\{ sa_{t_i} | \alpha \leq t_i \leq \omega, \forall i \right\} \quad (5.3)$$

5.4.2 *Time Window Manipulation*

A number of operations can be performed on the time window model, namely sizing, activation, deactivation, sliding, shrinking, expansion and overlapping.

Sizing: This sets the initial length of the time window. To determine the length of time windows, let the length of the initial time window, Ω_α^0, be set to w_0. The length of each time window is delimited by a minimum size, w_{min} , and a maximum size, w_{max}. The values of w_{min} and w_{max} are obtained from activity duration information that is derived from prior domain knowledge. For instance, w_{min} is set to the duration of the shortest activity, while w_{max} is set to that of the longest activity plus some slack time. Typically, given the initial time window, Ω_α^0, then the length of any new window, Ω_α^i, can be assigned using the formula below:

$$w_i = \rho * w_0, w_{min} \leq w_i \leq w_{max}, i = 1, 2, \ldots n \quad (5.4)$$

The value of ρ is chosen such that the resulting time window length lies between w_{min} and w_{max}. To minimize computational complexity, both w_{min} and w_{max} are set constant for all time windows. In this chapter, the value of $\rho = 1$ is chosen to set the default size of all time windows as equivalent to the initial time window. However, the default size is dynamically varied through the shrinking or expansion operations.

Activation/Deactivation: A Boolean flag, activated, is used to activate or deactivate a time window. It is set to true to indicate that the time window is active and false to show deactivation. By default, the flag is set to false and must be changed to true so as to use the time window model to segment a sensor data stream. Before the deactivation operation, the state of the time window must be logged in a suitable storage.

Sliding: The sliding operator allows the shifting of the current time window by some factor in order to derive a new time window. To determine the criteria for sliding, a sliding factor, μ, is defined. The sliding factor is a value that satisfies the constraint $0 < \mu \leq 1$. In this way we can obtain the size of the slide that needs to be applied to the current time window. The slide determines by how much the start time for the current time window, Ω_{α_i}, is shifted forward to determine the start time of the new time window, $\Omega_{\alpha_{i+1}}$. Let the slide to be applied to the current window to obtain the start time for the next time window be denoted by φ_i. We can compute φ_i by using the following formula:

$$\varphi_i = \mu * w_i, i = 0, 1, 2, \ldots, n \ldots \tag{5.5}$$

Given that the start time for the current time window is denoted by α_i and that of the succeeding time window by α_{i+1}, we can apply the slide to derive α_{i+1} using the formula:

$$\alpha_{i+1} = \alpha_i + \varphi_i \tag{5.6}$$

Overlapping: This refers to the process of having two or more-time windows active at the same time. By choosing a sliding factor value less than one ($\mu < 1$) two-time windows are made to overlap. A value of one ($\mu = 1$) means that the time windows are successive and non-overlapping. Furthermore, by examining the time windows being created and activated we can identify two properties:

Property (1): Two time windows, Ω_{α_i} and Ω^j_α, $i < j$, and Ω_{α_i} starts before Ω_{α_j}, are said to overlap in time if the start time, α_j, of Ω_{α_j} is less than the end time, ω_i, for Ω_{α_i}. This is denoted by the expression below:

$$\alpha_j < \omega_i \tag{5.7}$$

Property (2): Two time windows, Ω_{α_i} and Ω^j_α, $i < j$, are said to overlap in activations if property (1) is true and the intersection between the data sets in the two time windows is non-empty, i.e., at least one sensor activation belongs to both time windows. The non-emptiness is denoted by the following expression:

$$\overline{\overline{\Omega}}_{\alpha_i} \cap \overline{\overline{\Omega}}_{\alpha_j} \neq \emptyset \tag{5.8}$$

Whenever property (2) is satisfied, sensor activations can be used in two or more time windows during activity inference. This scenario can be used to facilitate the recognition based on complex activity patterns such as interleaved and concurrent activities.

Shrinking/Expansion: Given that the time window size is present, an ADL may be identified before the expiry of the window. Whenever this happens, the time window length may be reduced dynamically (this is called shrinking). Conversely, whenever it can be established that the time window may expire before an ongoing ADL is conclusively identified, the window length can be increased (this is called expansion) to keep the time window active for a little longer. To perform the shrinking operation and to compute the new window length, w_i', we define the shrink time, st. Shrink time, dynamically derived at runtime, refers to the time at which the decision to shrink the current time window is made. The new window length, w_i', is then computed using the formula below:

$$w_i' = st - \alpha_i \tag{5.9}$$

Similarly, to expand we define the length of expansion, exp. The new window length is then computed using the formula below:

$$w_i' = w_i + exp \tag{5.10}$$

5.5 Real-Time Data Segmentation for Continuous Activity Recognition

Once sensor activations are received, then by using the ADL ontology, each of them is converted into the corresponding ADL property assertion and added to a set of property assertions. At an appropriate time, the reasoning engine attempts to recognize the ongoing ADL. There are three modes that can be used to trigger reasoning. In the first mode, $\gamma = 0$ and each time a sensor is activated, activity recognition is performed. Using the second mode $\gamma = 1$ and reasoning occurs periodically at regular intervals during the length of the time window. The intervals can be set at configuration time. The activity inference engine should check the existence of new sensor activations before further attempts at reasoning. This requires that the current and previous sensor states are tracked to determine whether or not fresh activations have been obtained. Finally, using the third mode $\gamma = 2$ and reasoning occurs only at the expiry of the time window. The success of this mode depends entirely on optimal choice of time window lengths. At the deactivation of each time window, all sensor activations used within it are discarded.

5.5.1 Recognition Algorithms

In order to support continuous, real-time activity recognition we modify the algorithm discussed in Chaps. 3 and 4 [1, 12]. To manipulate the time window data structure the ontology-based activity recognition algorithm is enhanced with temporal segmentation ability as shown in Tables 5.1 and 5.2. Table 5.1 shows the pseudo-code for the time-window based algorithm, and Table 5.2 shows the pseudo-code for the ontological reasoning component of the algorithm.

Three operation modes are proposed to support the manipulation of time windows and to demonstrate the impact of dynamic manipulation. The first is no-shrink-no-expand, whereby the time window is effectively static and the size is not reduced or extended at runtime. The second is shrink-only mode for which the length of a time window can be reduced but cannot be extended. Finally, in shrink-and-expand mode, the length of a time window can be extended, reduced or both. The shrink-only and shrink-and-expand modes show the use of dynamic time windows. The mode used can be specified at configuration time. The pseudo-code in Table 5.2 supports both shrinking and expansion whenever shrink-only or shrink-and-expand modes are selected.

Table 5.1 Time-window segmentation based ontological activity recognition algorithm

```
Input: Receives the time window data structure (Ωα) and the ADL ontology (ADL-O)
Output: A matrix of time windows and corresponding text strings representing the
  likely activity labels, V.
RECOGNIZE-ADL (Ωα, ADL-O)
Set the initial time window
While active Do
   If initial window Then Activate time window End
   While time window unexpired Do
    Obtain and add sensor activations to Ωα
    If overlapping=true And overlapping window not active Then
         Compute slide and derive next time window
         Activate newly derived window
    End
    If reasoning mode = on-sensor Or at-intervals Then
         DO-ONTOLOGICAL-AR(Ωα,ADL-O)
    Else If reasoning mode = at-intervals And interval-elapsed
       Then
          DO-ONTOLOGICAL-AR(Ωα,ADL-O)
          End
   End(inner loop)
   If reasoning mode = on-expiry Then
        DO-ONTOLOGICAL-AR(Ωα,ADL-O)
   End
   Update matrix (V)
   Discard previous time window's sensor activations
   If overlapping=false Then
      Derive new window
      Activate time window
   End
End (outer loop)
Return (V)
```

Table 5.2 Ontological reasoning algorithm (adapted from [4])

```
Input: Receives the time window data structure (Ωα) with sensor
activations and the ADL ontology (ADL-O)
Output: A vector of text strings representing the likely
activity labels, A⃗.
```

```
  DO-ONTOLOGICAL-AR (Ωα, ADL-O)
  Derive property assertions
  Derive activity description, A_desc
  Perform equivalency and subsumption reasoning with A_desc to
determine underlying activity (ies), activity-1, activity-2,..,
activity-n
  If only one activity, e.g. activity-n, is obtained Then
    Check whether activity-n is abstract or specific
    If activity-n is specific Then
      Report activity-n is successfully identified
      If shrinkable Or shrinkable-and-expandable Then
         ATTEMPT-SHRINK (Ωα, ADL − O, activity − n)
         If window is not shrunk Then
            Report that more sensor data are needed
            ATTEMPT-EXPANSION (Ωα, ADL − O, activity − n)
         End
    End
    Add the activity-n label to vector, A⃗
    Else
     Report that more sensor data are needed
      If shrinkable-and-expandable Then
         ATTEMPT-EXPANSION (Ωα, ADL − O, activity − n)
      End
  Else
    Add all activity labels to vector, A⃗
    If shrinkable-and-expandable Then
       ATTEMPT-EXPANSION (Ωα)
    End
  End
  Return (A⃗)
```

5.5.2 *The Algorithm for Shrinking Time Window*

A time window can be shrunk under two conditions. Firstly, if all property assertions needed to describe a leaf activity have been specified, then the recognition system can truncate the current time window and spawn a new window. This is done by checking the activity description derived from the time window against the restrictions that have been defined for the given ADL class in the ontology. Secondly, the recognition system can choose to truncate the current time window if it determines that the ongoing activity has exhausted its duration and hence it is least likely to generate further sensor activations. Both cases have been captured by the pseudo-code in Table 5.2.

Table 5.3 Listing to shrink a time window

```
Input: Receives the time window data structure (Ωα) with sensor activations, the
ADL ontology (ADL-O), and the ADL ontology (activity-n)
Output: A truncated time window Ω'α
ATTEMPT-SHRINK (Ωα, ADL-O, activity-n)
If all properties of activity-n are asserted Then
   Shrink the window
Else If activity-n duration exhausted Then
   Shrink the window
Else
   If time-needed-to-complete-activity-n >= pending-window-length Then
      ATTEMPT-EXPANSION (Ωα, ADL-O, activity-n)
   End
End
Return (Ω'α)
```

Table 5.4 Listing to expand a time window

```
Input: Receives the time window data structure (Ωα) with sensor activations, the
ADL ontology (ADL-O), and activity label (activity-n)
Output: An expanded time window Ω'α
ATTEMPT-EXPANSION (Ωα, ADL-O, activity-n)
If activity-n is specific Then
 If some properties of activity-n are missing And needed time to complete ADL >=
      pending time window length Then
      Expand the window
 End
Else
   If some activations have been obtained Then
      Obtain subclasses of activity-n
      Derive maximum duration from the durations of these subclasses
    Obtain remaining duration to complete longest subclass, remainderADL
      If remainderADL >= pending time window length Then
         Expand the window
      End
   End
End
Return (Ω'α)
```

5.5.3 *The Algorithm for Expanding Time Window*

A time window can be expanded under two conditions. Firstly, given that a leaf activity has already been identified but the pending time window length is inadequate to complete the activity description, the window is expanded to allow additional activated sensors to be obtained (Table 5.3). Secondly, given that a leaf activity has not been identified, information about the currently identified generic activity is used to determine how much additional time would be needed to recognise its subclass that has the longest duration. The pseudo-code in Table 5.4 depicts this scenario.

5.6 An Example Case Study

The proposed approach has been developed in a SH-based activity recognition system. The system is implemented using Java language and a raft of semantic technologies and tools. Specifically, we developed ADL ontologies based on OWL-DL

[13] using Protégé editor [14] as shown in Fig. 5.3. The ADL ontology captures information about ADLs such as ADL concepts, hierarchical relationships among concepts, property restrictions for ADLs and contextual information, and sensor related concepts.

To support ontological reasoning, we have used Pellet [15] OWL reasoner, accessed through application programming interfaces (APIs) in Java, to provide reasoning capabilities for activity inference. In addition, we implemented the time window-based segmentation model as part of an activity recognition module. Figure 5.4 shows four system interfaces of the implemented system. Firstly, on the top-left it shows the interface that displays the set of all sensors that are currently deployed in the environment. Secondly, the top-right of Fig. 5.4 displays the configuration window that is used to add sensors to the SH environment, set the initial time window parameters, and to initialize the activity monitoring task. Thirdly, the dialog for choosing whether to monitor sensor activations in real-time or to play them back from a file is shown on the bottom-left. Finally, the list of all sensors that have been activated during a particular time window as well as the status of activity recognition is provided at the bottom-right of Fig. 5.4.

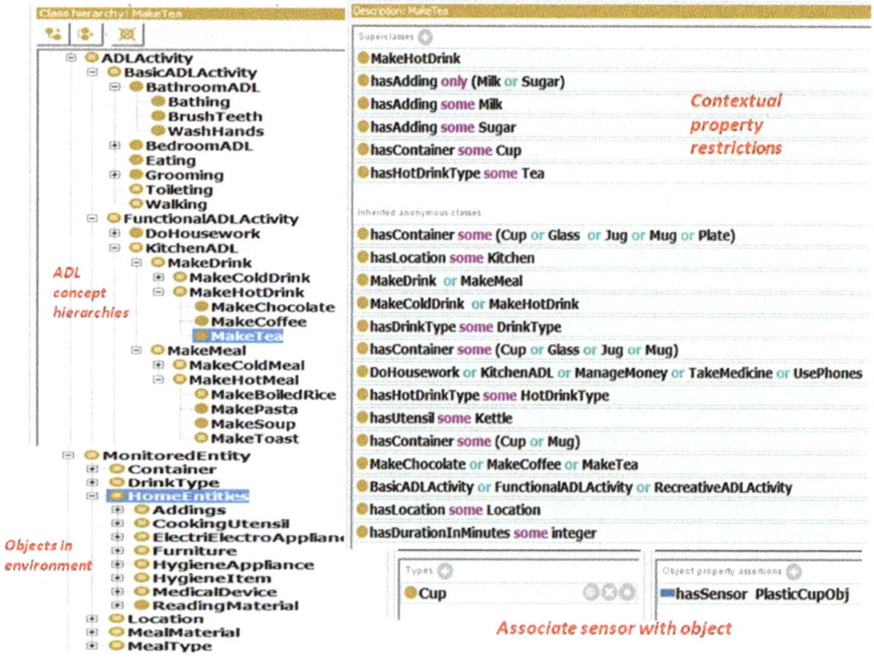

Fig. 5.3 Fragment of ADL ontology

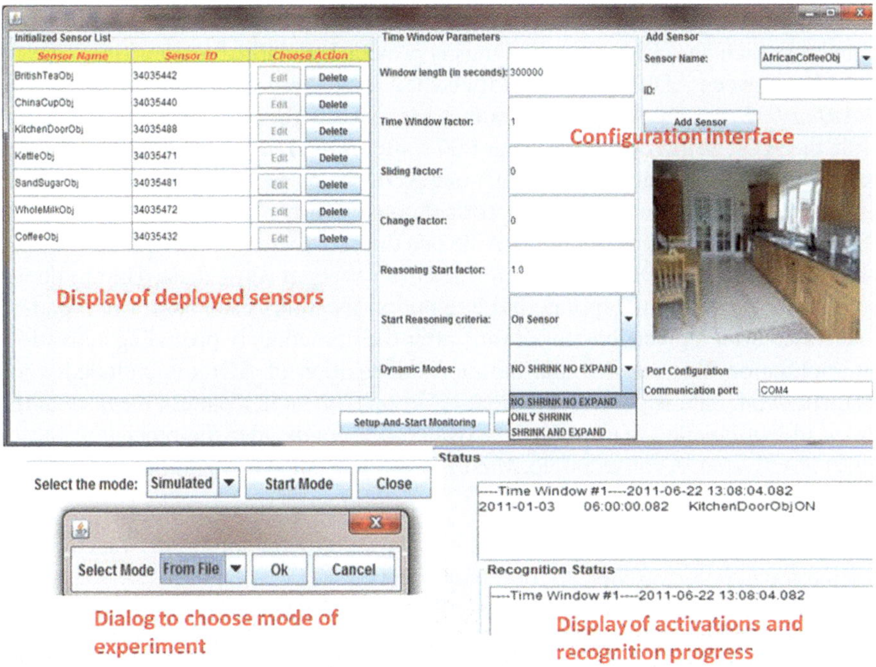

Fig. 5.4 System configuration and status display interfaces

5.6.1 *Experiment Design*

To evaluate and demonstrate the feasibility of the proposed approach, we developed a synthetic data generator that can be used to generate synthetic ADL data. This provides ADL data items that possess the necessary temporal information and allows us to quickly evaluate the feasibility of the developed approach before deployment in a real-world environment. Another advantage is that we are able to test the approach on different datasets. To facilitate data generation, we specified ‘seed’ ADL patterns at the start. Each seed ADL pattern is described by a sequence of ADLs. The synthetic ADL data generator then derives different permutations of these patterns. To select the permutation to use, it uses a random number generator. In this way, each permutation is given an equal chance of being considered an ADL pattern during dataset generation.

To generate the synthetic ADL data, eight typical ADLs related to meals (e.g. *MakeTea*, *MakeCoffee*, *MakeChocolate*, and *MakePasta*), hygiene (e.g. *HaveBath*, *BrushTeeth*, *WashHands*) and recreation (e.g. *WatchTelevision*) were used. In addition, to ensure that data is rich with useful temporal information, synthetic ADL data is generated corresponding to three-time periods per day: morning (6 am–9 am), midday (12 pm–2 pm), and evening (6 pm–10 pm). To facilitate this, the time period to which the ADL can be performed is specified when adding possible seed ADLs to the synthetic data generator. Similarly, when specifying seed patterns, the time period to which an ADL pattern belongs is provided. Finally, the transition time (in seconds) between ADLs is specified for each ADL pattern. For example, the pattern *MakeTea-0, BrushTeeth-600*, implies that MakeTea is the first ADL in the pattern, while the ADL *BrushTeeth* will occur 600 s after *MakeTea* is completed. Currently, we have made the assumption that only one ADL can be performed at the same time. As a result, no two sensors can be activated concurrently during data generation.

For each ADL considered, we provide one or more patterns of sensor activations. Given that the same ADL may be performed in a variety of ways; these patterns depict the various ways. To incorporate more temporal meaning, each sensor in a pattern is activated after a given amount of time after the immediately preceding activation. By implication, this ensures that duration information of ADLs is included when synthetic ADL data is generated. The text *SensorObj@n* in a pattern means that the sensor object labelled *SensorObj* is activated n seconds after the preceding sensor object is activated. As an example, *MakeTea* is represented by the following patterns of activated sensors.

- **Pattern#1**: KitchenDoorObj@0, KettleObj@20, CupObj@180, TeaObj@20, MilkObj@20, SugarObj@20 (duration = 260 s)
- **Pattern#2**: KitchenDoorObj@0, KettleObj@20, CupObj@180, TeaObj@20, MilkObj@20 (duration = 240 s)
- **Pattern#3**: KitchenDoorObj@0, KettleObj@20, CupObj@180, MilkObj@20, TeaObj@20, SugarObj@20 (duration = 260 s)

Each sensor pattern is captured in a collection data structure; with a collection available for each ADL. Similar sensor activation patterns are specified for the other ADLs too. To select the sensor pattern for any ADL, the synthetic data generator uses a random number generator to randomly select the index of the sensor pattern for the relevant ADL from the relevant collection. This eliminates bias and gives each pattern a fair chance of being selected.

To test the time window approach and associated algorithms, a simulation tool has been built to mimic the activation of sensors in a dense sensor-based deployment. The simulation tool plays back the synthetic ADL data generated described above and feeds the sensor data to the activity recognition system as if the sensor activation is occurring in real-time. As the data is played back, the recognition engine tries to identify the ongoing ADL and displays the status on the interface shown at the bottom-right of Fig. 5.4.

5.6.2 Time-Window Model Configuration

To carry out the experiments, the duration of the default time window is initially set to a value slightly greater than the longest ADL - in the experiments the longest ADL by duration is MakePasta. The reasoning start mode (γ) is set to zero (0) so that activity inference is attempted each time sensor activation is obtained. The sliding factor (μ)

is set to one (1) to indicate that time windows are consecutive and non-overlapping. The time window factor (ρ) is set to one (1) so that each time window is by default the same size as the initial window. Finally, the change factor (δ) is computed at runtime during shrinking and expansion operations. Similarly, the other parameter (i.e. start time (α), end time (ω), sensor data set ($\overline{\overline{\Omega}}_{\alpha}$) and the vector of activity labels ($\vec{A}$) are dynamically computed at runtime.

5.6.3 Ground-True Synthetic ADL Data

To facilitate analysis, we generated synthetic ADL data for four weeks based on the eight ADLs above. The dataset contains a total of 154 ADL activities and a summary of the ADLs is provided in Table 5.5. Three variables are used to describe the dataset. These are: (1) %in-pattern ADLs- describes proportion of the ADL that appear in ADL patterns that have at least two ADLs; (2) %standalone ADLs- describes the proportion of the ADL that appear in single-ADL patterns and; (3) the total number of times a given ADL occurs in the dataset. Generally, ADLs that participate in many ADL patterns (i.e., MakeTea,MakePasta, BrushTeeth, HaveBath and WatchTelevision) have more instances. Conversely, those that appear in just one ADL pattern (i.e., MakeCoffee, MakeChocolate and WashHands) typically have just a few instances. Using the real-time activity recognition system, we played back these synthetic ADL data and present the results in the next section.

Table 5.5 Summary of synthetic ADL datasets

ADL name	Description of dataset		
	% Standalone ADLs	% In–pattern ADLs	Total instances
MakeTea	10	90	41
MakeCoffee	100	0	4
MakeChocolate	100	0	3
MakePasta	10	90	29
BrushTeeth	30	70	19
HaveBath	10	90	28
WashHands	100	0	6
WatchTelevision	10	90	24
Num. Of ADLs			154

5.6.4 Experiment Result Analysis

In order to evaluate the performance and therefore the feasibility of the approach, we used the metric accuracy. Accuracy measures the correctness of the algorithm, i.e. the ability of the algorithm to return correct results. We compute the accuracy of the recognition performance and provide the results in Table 5.6. The accuracy is computed from the values of true positive (tp), false positive (fp), true negative (tn), and false negative (fn) using the formula:

$$\text{accuracy} = \frac{\text{tp} + \text{tn}}{\text{tp} + \text{fp} + \text{tn} + \text{fn}}$$

We report results for three experiments and the first experiment was to evaluate recognition performance for static time windows, i.e., given that time windows cannot be shrunk or expanded. The results are shown in Table 5.6. The second experiment evaluated recognition performance when shrink-only is enabled. The results are presented in Table 5.7. Finally, the third experiment evaluated the performance given that shrink-and-expand is selected. The results are shown in Table 5.8. The second and third experiments relate to dynamically manipulated time windows.

5.6.5 Findings and Discussions

As can be seen in Table 5.6, the recognition accuracy of *MakeTea*, *MakePasta*, *MakeCoffee*, *MakeChocolate* and *WashHands* is quite encouraging and attests to the feasibility of the presented approach. However, it is important to note that the recognition accuracy for *BrushTeeth*, *HaveBath* and *WatchTelevision* is low compared to the other ADLs. This can be attributed to the fact that these ADLs occur in very few standalone patterns; instead they mostly appear in sequential patterns. Given that

Table 5.6 Recognition accuracy without shrinking or expansion

ADL	Values from dataset				
	TP	FP	TN	FN	Accuracy
MakeTea	39	0	0	2	0.951
MakeCoffee	4	0	0	0	1.000
MakeChocolate	3	0	0	0	1.000
MakePasta	28	0	0	1	0.966
BrushTeeth	14	0	0	5	0.737
HaveBath	19	0	0	9	0.679
WashHands	6	0	0	0	1.000
WatchTelevision	10	0	0	14	0.417
Average accuracy					0.844

Table 5.7 Recognition accuracy with only shrinking enabled

ADL	Values from dataset				
	TP	FP	TN	FN	Accuracy
MakeTea	39	0	0	2	0.951
MakeCoffee	4	0	0	0	1.000
MakeChocolate	3	0	0	0	1.000
MakePasta	26	0	0	3	0.897
BrushTeeth	17	0	0	2	0.895
HaveBath	24	0	0	4	0.857
WashHands	6	0	0	0	1.000
WatchTelevision	18	0	0	6	0.750
Average accuracy					0.919

Table 5.8 Recognition accuracy with both shrinking and expansion enabled

ADL	Values from dataset				
	TP	FP	TN	FN	Accuracy
MakeTea	39	0	0	2	0.951
MakeCoffee	4	0	0	0	1.000
MakeChocolate	3	0	0	0	1.000
MakePasta	21	0	0	8	0.724
BrushTeeth	17	0	0	2	0.895
HaveBath	24	0	0	4	0.857
WashHands	6	0	0	0	1.000
WatchTelevision	16	0	0	8	0.667
Average accuracy					0.887

the time window does not dynamically vary once created, sensor data belonging to more than one ADL in the pattern may be merged within a time window, thus leading to poor recognition accuracy.

Results in Table 5.8 show that there is a significant improvement on overall recognition accuracy. However, compared to Table 5.6, there is a reduction in the accuracy for *MakePasta* and a corresponding increase for *BrushTeeth*, *HaveBath* and *WatchTelevision*. Similarly, results in Table 5.7 indicate that the overall recognition accuracy was highest compared to the Tables 5.6 and 5.8. Just like in Table 5.8, the recognition accuracy of *MakePasta* reduced while that of *BrushTeeth*, *HaveBath* and *WatchTelevision* increased. However, despite the reduced recognition accuracy observed for *MakePast*a in both in Tables 5.7 and 5.8, there is an overall improved average accuracy. The direct comparison of recognition accuracy per ADL is shown in Fig. 5.5. In addition, Fig. 5.6 shows a direct comparison of average recognition accuracy.

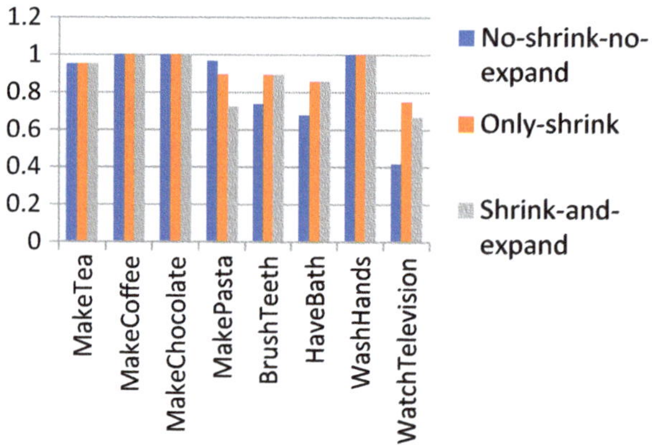

Fig. 5.5 Comparison of recognition accuracy per activity

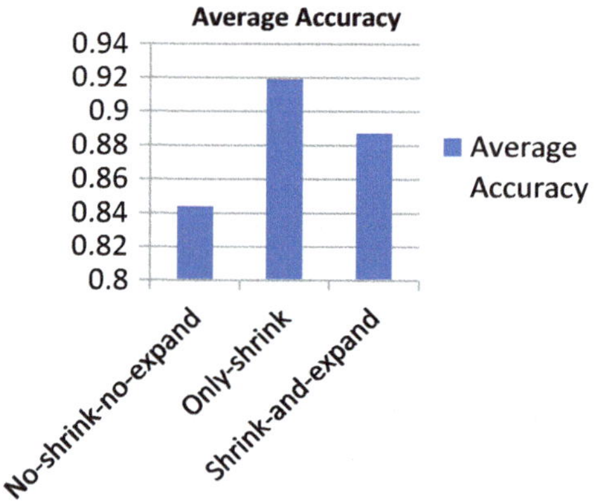

Fig. 5.6 Comparison of average recognition accuracy

In all experiments, the activity recognition system recognized all the instances of all standalone ADLs. The reduced recognition accuracy was only observed regarding the in-pattern instances. As a result, we have reason to believe that the performance of activity recognition regarding *BrusthTeeth*, *HaveBath* and *WatchTelevision* may have been affected by the fact that they occur with other ADLs, and more so as ubsequent ADLs in the ADL patterns. In addition, the transition times from one ADL to another in an ADL pattern could also have made it possible for sensor data belonging to two distinct ADLs to be aggregated into one description, resulting in non-recognition. Another reason is the fact that several variants of ADLs were generated in the dataset. Whenever shrinking was allowing, a window could be shrunk before all the sensor activations associated with an ADL are obtained, hence the activations that arrive later could be merged with subsequent activations thus causing recognition failures.

In the no-shrink-no-expand case, the recognition failure could be attributed to the chosen time window sizes.

We believe that by handling transitions between adjacent activities the recognition accuracy can be improved. This explains better recognition accuracy when shrinking and expansion are allowed. However, to minimize the failures whenever shrinking and expansion are supported, we believe that explicit relationships between ADLs in a pattern should be defined. This should involve a characterization of additional temporal relationships for the ADLs that could occur in patterns.

An interesting finding from the comparison of experiment results in Table 5.3 is that shrink-only had the best recognition accuracy. The shrink-and-expand case is the next best performer. We attribute lower recognition accuracy of shrink-and-expand compared to shrink-only to the fact that the maximum duration was used to derive time window lengths, thereby favouring activities with longer duration at the expense of those with shorter durations. However, we believe that by incorporating information about how inhabitants perform activities that can be captured through continuous use will improve the recognition accuracy of both shrink-only and shrink-and-expand mode. In future we plan to integrate the activity learning and model evolution approaches described in [16] so that feedback from how an inhabitant performs tasks is used by the recognition system for adaptation.

The average accuracy for all the experiments is above 83% using the provided dataset as well as with other datasets that we generated. This demonstrates that the approach is feasible in supporting real-time activity recognition.

5.7 Summary

This chapter presented an approach based on dynamically varied time windows to support sensor data segmentation for use in continuous, real-time activity recognition. It characterises activity recognition and sensor data segmentation from which it formally defines a time window-based segmentation model. The chapter has detailed the rationale and operation algorithms of the model in the context of knowledge-driven activity recognition. In addition, different scenarios regarding dynamic manipulation of time windows were discussed. The model allows rich temporal information associated with sensor data and activities to be exploited in real-time activity recognition.

As a case study to illustrate the approach, an implementation of a prototype to evaluate the approach was also described. The prototype consists of a synthetic ADL data generator, ADL ontology, sensor data simulator for ADL data playback, and a real-time activity recognition system. To establish the feasibility of this approach, this chapter has presented evaluation results from three experiments. Accuracy was chosen as an evaluation metric and the resulting average accuracy has demonstrated the feasibility of the approach. The average recognition accuracy was lowest, at 84%, when no-shrink-no-expand mode was activated. It was highest, at over 91%, when shrink-only mode was enabled.

References

1. Chen L, Nugent C (2009) Ontology-based activity recognition in intelligent pervasive environments. Int J Web Inf Syst 5(4):410–430
2. Riboni D, Pareschi L, Radaelli L, Bettini C (2011) Is ontology-based activity recognition really effective? In: 2011 IEEE international conference on pervasive computing and communications workshops, PERCOM workshops 2011
3. van Kasteren T, Noulas A, Englebienne G, Kröse B (2008) Accurate activity recognition in a home setting. In: Proceedings of the 10th international conference on Ubiquitous computing, UbiComp'08
4. Akdemir U, Turaga P, Chellappa R (2008) An ontology based approach for activity recognition from video. In: Proceeding of the 16th ACM international conference on multimedia, MM'08
5. Bao L, Intille SS (2004) Activity recognition from user-annotated acceleration data. In: Ferscha A, Mattern F (eds) Pervasive Computing. Springer, Berlin, pp 1–17
6. Huỳnh T, Blanke U, Schiele B (2007) Scalable recognition of daily activities with wearable sensors. In: Hightower J, Schiele B, Strang T (eds) Location- and context-awareness. Springer, Berlin, pp 50–67
7. Tapia EM, Intille SS, Larson K (2004) Activity recognition in the home using simple and ubiquitous sensors. In: Ferscha A, Mattern F (eds) Pervasive computing. Springer, Berlin, pp 158–175
8. Hong X, Nugent CD (2009) Partitioning time series sensor data for activity recognition. In: 2009 9th international conference on information technology and applications in biomedicine, pp 1–4
9. Ortiz Laguna J, Olaya AG, Borrajo D (2011) A dynamic sliding window approach for activity recognition. In: Lecture notes in computer science (including subseries Lecture notes in artificial intelligence and lecture notes in bioinformatics)
10. van Kasteren T, Noulas A, Englebienne G, Kröse B (2008) Accurate activity recognition in a home setting. In Proceedings of the 10th international conference on Ubiquitous computing, UbiComp'08, pp 1–9
11. Hong X, Nugent C, Mulvenna M, McClean S, Scotney B, Devlin S (2009) Evidential fusion of sensor data for activity recognition in smart homes. Pervasive Mob Comput
12. Chen L, Nugent CD, Wang H (2012) A knowledge-driven approach to activity recognition in smart homes. IEEE Trans Knowl Data Eng 24(6):961–974
13. Horrocks I (2005) OWL: A description logic based ontology language. In International conference on principles and practice of constraint programming, pp 5–8
14. Stanford Center for Biomedical Informatics Research (2011) The protege ontology editor and knowledge acquisition system. Available from: https://protege.stanford.edu/
15. Stardog-union: Pellet: OWL 2 Reasoner for Java. https://github.com/stardog-union/pellet
16. Okeyo G, Chen L, Wang H, Sterritt R (2010) Ontology-enabled activity learning and model evolution in smart homes. In: Lecture notes in computer science (including subseries Lecture notes in artificial intelligence and lecture notes in bioinformatics)

Chapter 6
Semantic-Based Sensor Data Segmentation

6.1 Introduction

Activity recognition can be conceptualised in five-phases as depicted in Fig. 6.1. The segmentation phase is responsible for organising the observed sensor events based when a single inhabitant performs one or more activities in a sequential or interweaving scenario. In order to make segmentation decisions, prior knowledge model is required to verify association links such as what everyday object is the sensor attached to, contextual information (i.e., location, time and ambient attributes) of the object and what ADL(s) is this object used for. The data from the set of segmented sensor observations for a given activity is later analysed by the AR algorithms to determine whether the actions were completed with a satisfactory evidence (i.e., if the cooker knob rotated to low, medium, high or off state) and provide effective assistance when necessary. Therefore, a correctly segmented set of sensors can boast AR algorithm accuracy, performance and reduces computational resources being wasted on irrelevant sensor data.

Recent studies have applied time series (discussed in Chap. 5), statistical and probabilistic [1] analysis approaches to sensor data segmentation, which failed to separate sensor observations based on the relation to ongoing activities in real time. This chapter explores the relationships and metadata of sensor data to activities to segment unfolding sensor events into relevant set of activities of daily living (ADL). In addition, it also presents methods on how individual preferences can be incorporated within the semantic segmentation process.

Knowledge-driven (KD) approach has received an increasing amount of interest to express complex relationships between sensors and domain-specific knowledge. The process of defining complex sets of relationships has been investigated in the past studies and they can be categorised as syntactical, semantic and pragmatic in information theory [2]. In syntactical approach, a concept represented in a two or more non-syntactically equivalent statements are assumed to be statements of independent concepts. In contrary, the semantic approach is concerned about representing the meaning of a concept using relationships [2, 3], hence, the same concept can

L. Chen and C. D. Nugent, *Human Activity Recognition and Behaviour Analysis*, https://doi.org/10.1007/978-3-030-19408-6_6

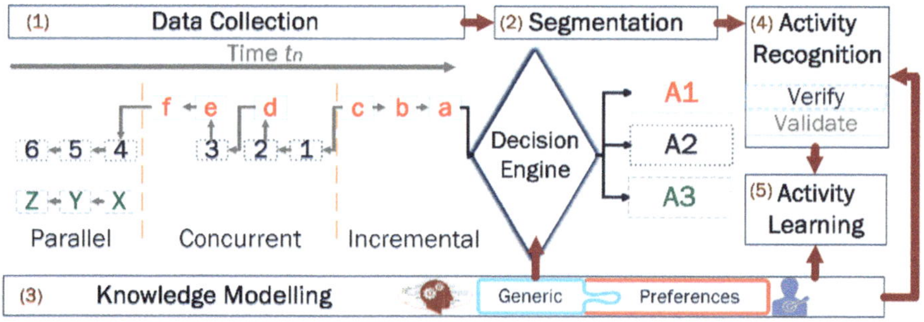

Fig. 6.1 Illustrating five interdependent phases of activity recognition

be syntactically represented in more than one statements but mean the same thing. The pragmatics approach studies the relations between a concept and inhabitant in a given context of interest. The benefit of adapting syntactical approach is that knowledge can be structured using defined syntax, queried and interpreted by the machine, however, suffer from the flexibility of expressing intricacy of relationships and meaning between two concepts that pragmatic and semantic approaches can provide. The semantic theory has its roots from semiotics in philosophy which in general is a study of signs and its significations (meaning) [4]. These signs can be words, images, sounds, gestures and objects. Hence, the semantic theory is studied heavily in cognitive philosophy, natural language and machine learning [5]. The following sections highlight recent studies proposed to segment sensor events that adapt the aforementioned notions of the three information theories.

6.1.1 *Semantic Approach: Indirect Query and Rules*

Work in [6, 7] adopted ontological models to describe ADLs, environmental entities and their relations along with other methods to classify and infer unfolding activities. However, they do not directly inspect each arrived sensor event and then segment to the appropriate queue related to ongoing activities. Instead, the continuous queries or rules are executed on events stored in the database and knowledge model without using any automatic reasoners to determine the relationship between events and ADLs. Similarly, work in [8] proposed C-SPARQL, an extension to SPARQL Protocol and RDF Query Language (SPARQL) where individual sensor events in a stream are annotated with a timestamp and continuously queried using a specific window size. The key limitations of the approach are the classical multi-query optimisation problem where the challenge is to identify the common parts, adapting/reformulating the order in which queries are executed with the ability to dynamically change the window size. Another stem of work, [9, 10], used Semantic Web Rule Language (SWRL) based inferencing rules to define the nature of activities with a temporal representation technique. These SWRL rules and Java Expert System Shell (JESS) rule engines were used to segment the sensor events using their timestamp information and perform entailments for the complexity of the ongoing activities. One of the major limitations of this approach is that an attempt to use generic ontology reasoner is made, however, it is unclear if reclassification of the whole ontology is done incrementally or not. In the case of the non-incremental reclassification approach, the performance and scalability can degrade exponentially as the size of an ontological model and data grows. Furthermore, rules can be generated for general purpose and also for inhabitant specific preferences as provided in the study in [11]. However, each time the new rules are added or updated to enrich the knowledge base (KB), the whole ontological model is reclassified. In addition, managing models generated using generic and inhabitant specific rules exclusively adds to the complexity further.

6.1.2 Syntactical Approach: RDBMS and Semantic KB Mapping

Similarly, work in [12] presents a layered ontology and complex event processing (CEP) engine based framework, namely, AALISABETH, to segment the sensor observations. The framework integrates temporal based reasoning with a dynamic time window sizing mechanism to segment the incoming data and perform AR in real-time. The approach leverages Esper solution for CEP and D2RQ engine to map data into RDF graphs. Although the framework utilises highly optimised, scalable Esper CEP engine solution and is open source, the system falls short in directly segmenting the incoming sensor data semantically in real-time as it arrives from the sensor network. This limits the client applications to receive an event-based notification which is critical in an emergency situation such as fall detection. Another key limitation of the framework is that the event data from the sensor network is stored directly into a traditional relational database management system (RDBMS) without inspecting individual events and segmenting them appropriately or appending to an ongoing activity queue. Instead, to filter or segment sensor events for a given ADL, continuous queries are required to be executed in order to obtain a set of sensor events between a specific time range/number of records and then perform Web Ontology Language (OWL) based reasoning capabilities to find any relevance to the activity of interest. Alternatively, the Pellet reasoner which has incremental reasoning support (i.e., only affected changes in the ontology are classified) could be further utilised instead of creating an overhead to query and map each of the events from the RDBMS database using the D2RQ tool. Furthermore, the framework is not intended to cater for inhabitant's preferences when performing a generic ADL.

6.1.3 Pragmatic Approach: Precondition and Evidential Theory

Work in [13] presents an event filtering approach by adding preconditions with probabilities on the phases when carrying out each ADL in order to segment the incoming events. It is unclear how the algorithm can detect new activity when an action is shared amongst more than one activities and it can either be part of a main activity or precondition actions for another activity. For instance, *MozzarellaCheese* can be part of the precondition of *MakePizza* ADL and postcondition for *MakeCheesyToast* ADL. This approach has achieved good accuracy in segmenting and recognising composite activities but there is the scope for improvement in terms of recognising other scenarios. Another work in [14] leveraged evidential theory and proposed three segmentation algorithms based on location, activity model and dominant-centred (key actions for a activity) for non-interleaved and interleaved activities. The location and activity model-based segmentation algorithms fall short in distinguishing activities when performed in the same location and with similar everyday objects for activities compared to the dominant algorithm. There is a little implementation detail provided by the authors, however, one of the key limitations of all the three algorithms is the lack of support for user preferences and a reasoner to automatically detect and recognise the activity.

This chapter presents five contributions: (i) a semantic-enabled segmentation approach which combines generic and personalised ADL knowledge that enables simple and composite ADLs to be recognised in real-time; (ii) a KB model capturing the relationships between entities in the house and ADLs; (iii) a pragmatic and light-weight mechanism to manage inhabitants specify preferences for conducting a given ADL; (iv) a semantic decision engine algorithm; (v) system implementation details and a prototype to evaluate the approach and present the findings.

6.2 Semantic-Based Approach to Sensor Data Segmentation

The semantic theory-based segmentation approach analyses relationships between sensor events and an everyday object and its significance as an action to a set of known ADLs. This will enable disentangling composite activities with actions performed in no particular order and organise them separately to allow further activity classification and learning tasks. A knowledge modelling building block is developed in Sect. 6.2.1 which conceptualises and captures the environmental context (i.e., ambient attributes, everyday objects, location, sensors), generic and inhabitant specific preferences to perform ADLs and their semantic relationships into an ontological model. A semantic decision engine is developed in Sect. 6.2.2 to make segmentation decisions based on three inputs: the new observed sensor event, the ontological model and a set of previously segmented sensors for a given activity. A notion of multithreading is adapted to separate tasks of buffering sensor data stream, event recycling, decision engine, managing ADL threads and manipulating data from the graph-based database. This multithreading mechanism to semantically segment sensor event is described with a pseudo algorithm in Sect. 6.2.3.

Figure 6.2 depicts the overall segmentation approach. As the sensor events are initially added to the data stream, multiple ADL threads, generic and preference, analyse the sensor events using decision engine and store the relevant events independently. Therefore, one sensor event can be shared between two different activity threads with different ADL goals. For instance, opening *Fridge* action detected by sensor *e* at T_n can be shared with *MakeTea* ADL and *MakePasta* ADL thread. The ADL threads manager creates a new ADL thread (*NEW_ACTIVITY*) only when the sensor event is not part of any ongoing ADL threads otherwise the event recycler thread updates the sensor data stream. There are two types of ADL threads being created to capture generic actions (sensor b attached to *PastaBag*), for a given activity (*MakePasta*), and if the observed event (sensor d attached to *HotSauce* at T_n) is part of the personalised actions for that activity (i.e., *PrefMakeVegPasta*). The decision engine determines if the new sensor event, along with the previous set of sensors for a given activity is part of the pre-defined generic set of actions by performing semantic reasoning and invoking queries to the TDB for personalised actions. The new preference thread (*NEW_PREF_THREAD*) is only created when the new sensor event is part of a personalised action for a given ongoing activity and there is no active preference thread. Moreover, each ongoing activity thread with the segmented set of sensors data will enable further validation of AR accuracy, timeout and completion procedures, i.e. storing relevant information and prompting the inhabitant when appropriate in future work.

6.2.1 Object, ADL and Context Relationships Modelling

The key building block of ADL modelling consists of three phases; (1) environmental context (EC) modelling, (2) semantic relationships (SR_i) modelling and (3) personalised ($Pref_j$) object interactions. In the first phase, the object-oriented notion (classes and instances) is adapted to conceptually describe the physical or metaphysical entities (ET_k) and their attributes as classes (C) to form an overall environmental context (EC) for a given smart home environment. The key entities considered are a person (X_n), rooms (Location, L_m) and ambient characteristic (AC_p), sensor characteristics (S_o) and everyday fixed/portable objects (Obj_x); see Eq. (6.1):

$$EC = X_n,\ L_m,\ AC_p,\ So,\ Obj_x \tag{6.1}$$

The second phase records semantic relationship (SR) properties between EC classes and ADLs. The instances of EC classes (i.e., everyday objects) are then created for sensor environment (SE) to create a relationship (R_e) between sensor

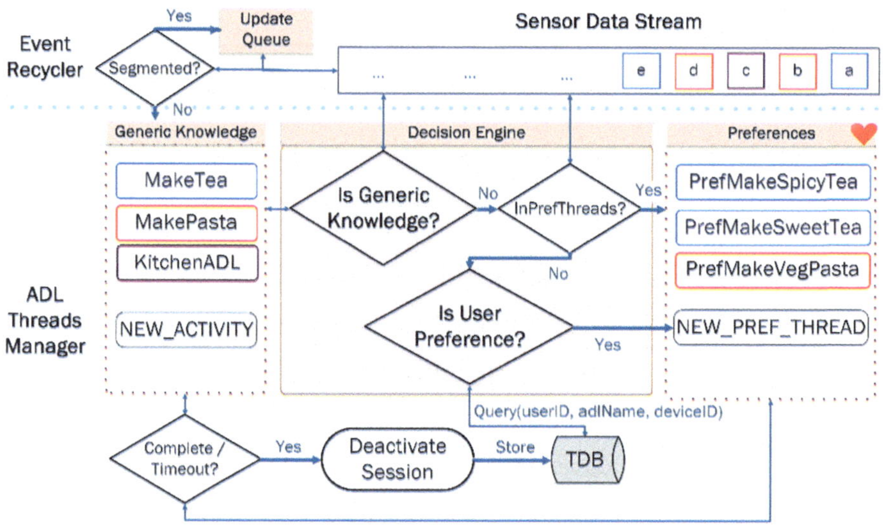

Fig. 6.2 Overview of the semantic segmentation approach with $\mathcal{T}$-box and $\mathcal{A}$-box KB

event, object it is attached to and this object's use in ADLs; see Eq. (6.2). This abstraction in ADL actions description encourages decoupling, reuse and adding the further meaning of the actions to the activity using R_e. For example, *MakeTeaADL* (subset of *MakeHotDrinkADL*) class describes the actions using *hasHotDrinkType* (R) relationship property with Tea (C) and the characteristics of the property are described to be only used for *MakeHotDrinkADL* (domain) and everyday objects that are used for *HotDrinkType* (range). This means if no other ADL that is a subset of *MakeHotDrinkADL* that has a *hasHotDrinkType* property with *Tea*, it can be deduced that this action is potentially a part of *MakeTeaADL*. Similarly, other actions for *MakeTeaADL* can be described using *hasUtensil*, *hasContainer* and *hasAddings* properties for using the kettle and adding sugar and milk to the teacup. Figure 6.3 show the relationships between a set of EC classes and *MakeTea* ADL to show the meaning of inhabitant's action.

Moreover, the sensor environment (SE) information is then encoded to describe existing set of EC items available in the given residential environment and the sensor attached to it as instances (I_w). Therefore, instances of EC (iEC_w) such as environmental objects ($iObj_w$) and sensor (iS_w) with their relevant classes (C_n) are explicitly described with the relationship (Re) between them initially. For example, to1 is an instance of *ContactSensor* (S) that *isAttachedTo* (R) a *RedKettleObj1* ($iObj_w$) which is a class type of *Kettle* (Obj_x). The observed values/states of an iS_w are stored as primitive data types (pt_u) for a single observation or creating another instance of an observation class containing the primitive data for multiple observations; see Eq. (6.3).

$$SR = ADL_n(R_e,\ EC_n) \rightarrow R_e \rightarrow SE; \tag{6.2}$$

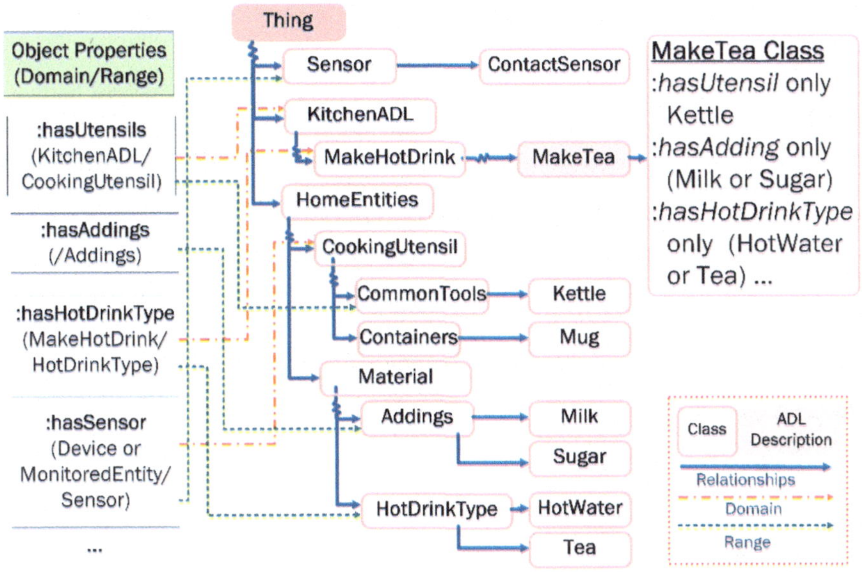

Fig. 6.3 Semantic relationship properties between *MakeTea* ADL, objects, and sensor characteristics

$$\mathrm{SE} = I_{\mathrm{w}}(\mathrm{R_e},\ \mathrm{S_o}) \rightarrow \mathrm{R_e} \rightarrow I_{\mathrm{w}}(\mathrm{R_e},\ \mathrm{ET_k})||I_{\mathrm{w}}\left(\mathrm{R_e},\ \left\{\mathrm{pt_u}\right\}\right) \tag{6.3}$$

The final phase is to capture inhabitant specific preferences ($\mathrm{Pref_j}$) that are subjective to individual's cultural background and rituals followed to carry out a given ADL. It is important to keep the generic (factual and commonly accepted by the wider community) and personalised sets of ADL description disjointed to avoid generalising or assuming both must be actioned to complete the activity. Therefore, instances that are members ($\mathrm{R_e}$) of *Preference* and $\mathrm{ADL_n}$ classes are created to capture actions or ambient attributes using $\mathrm{iEC_w}$ that are specific to a person ($\mathrm{X_s}$); see Eqs. (6.4) and (6.5). For example, an individual *Bob* (I) who is a type of *Male* (C) has set of instances of Preferences} that are linked with *hasPreference* relationship (R). An example of a preference instance is *BobMakeSpicyTeaPref* (*Pref*) which is a type of *Preference* (C) and *MakeTeaADL* (ADL) with a set of iEC instances, i.e., *GingerObj*(I) and *CinnamonObj*(I). This statement means that *Bob* has a preference to make tea and he may/like to put a ginger and cinnamon in his tea.

$$\mathrm{X_n} = I_{\mathrm{w}}(\mathrm{R_e},\ \mathrm{Human} \subseteq \mathrm{Male}) \rightarrow \mathrm{R_e} \rightarrow \left\{\mathrm{Pref_1},\ \ldots\ \mathrm{Pref_j}\right\} \tag{6.4}$$

$$\mathrm{Pref_j} = I_{\mathrm{w}}\left(\mathrm{R_e},\ \mathrm{ADL_n} \prod \mathrm{Preference}\right) \rightarrow \mathrm{R_e} \rightarrow \mathrm{I_w}(\mathrm{R_e},\ \mathrm{iEC_w}) \tag{6.5}$$

6.2.2 Semantic Decision Engine

The role of the semantic-based decision engine is to identify relationships between the sensor, everyday object and actions described in ADLs based on ontological model and triplestore querying. This allows decision engine to distinguish ADL actions occurring in any order for a single or multiple ADLs in a composite manner. The common ADL actions are automatically recognised using terminology box ($\mathcal{T}$-box) reasoning method with incremental Pellet reasoner and inhabitant specific actions using assertion box ($\mathcal{A}$-box) reasoning method. The decision engine is utilised by individual activity threads in order to find an association with new, previously observed events and candidate ADL class. The classification of candidate ADL class is continuously updated and refined with further evidence of actions that satisfies the ADL descriptions.

The decision engine takes three inputs, processes them into two stages and outputs the updated results. The three inputs are (1) semantic-based KB model created in Sect. 6.2.1, (2) activity thread (AT_n) attempting to find relations with the (3) new sensor event (e_m). Each AT_n contains structured information about generic and preferred actions observed as sensor events, ADL class and list of preferences matched that are associated to the inhabitant. The two-stage decision-making process updates the activity thread accordingly as the new sensor events are inspected incrementally for any association.

$$\begin{aligned} AT_n = \{&\text{tbox}[\text{class:someADL},\ s\{\ldots,\ e_m\}\}], \\ &\&\ \text{abox}\left[\text{Pref}_j\left[\text{name:somePref},\ s\{\ldots, e_m\}\}\right]\right]\} \end{aligned} \tag{6.6}$$

In the first stage of the decision-making process, generic semantic relationships are traced from EC to SR and SR to SE compared to inverse when developing the KB model [15]. Therefore, the metadata of a sensor observation e_m is analysed to find the ET the sensor is attached to and deduce the potential R_n with a set of ADL_n description. This metadata within KB consists relationship properties such as domain and range for a given ET. Therefore, the association between ET_k, (i.e., everyday objects) and ADLs can be automatically inferred using semantic reasoners or simply querying the KB model. This process is known as terminology box ($\mathcal{T}$-box) reasoning [16].

The second stage is only executed when the result returned from $\mathcal{T}$-box reasoning identifies any conflicts with the ADL class description. The conflicts can be raised when a given sensor attached to an ET is forced to be part of a given ADL which is outside the restricted set of ET_k. In this case, it is assumed that ET is part of inhabitant's preferences or part of a new set of actions for ADL_n. The preferences are currently pre-defined and stored as individuals containing a list of iEC_s that an inhabitant prefers to use to perform a given ADL. Therefore, semantic queries are made to extract all preferences of the inhabitant (*userID*) for a given ADL (*adlName*) that as sensor observation (*deviceID*) as an action. This process is known as assertion box ($\mathcal{A}$-box) reasoning.

The semantic reasoner carries out several tasks using $\mathcal{T}$-box and $\mathcal{A}$-box knowledge which includes but not limited to: satisfiability, subsumption, consistency checking equivalence, disjointness, and instance checking [15, 17]. The satisfiability task is to ensure the class description (axioms) is not contradictory. The subsumption task ensures class B satisfies all the inheriting properties (R) of parent class A. The consistency checking ensures classes and their instances do not violate the axioms descriptions. The instance checking ensures the relationships with other instances are within the boundary of a set of classes it can subsume. The equivalence task is to match the two concepts with respect to its properties in contrary to disjointness tasks. The conjunctive querying answering is performed at the second phase of decision engine to identify inhabitant's preferences with a given ET using relationships between instances of EC and ADLs.

Figure 6.4 illustrates the three inputs taken by the decision engine to verify if the new sensor observation Ginger(e_5) is part of the generic/personalised action of the ongoing *MakeTea* activity (AT1). Initially, a new activity thread, AT1, is created to add the first sensor observation, *Fridge* (e1), into the empty set of sensors and the results returned from two-stage reasoning process. In this case, e1 is inferred by the generic $\mathcal{T}$-box reasoner to be part of *KitchenADL* in the first stage of decision engine. As the new sensor event, e_2 occurs, the current AT1, temporarily add it to the list $\{e_1, e_2\}$ and perform the generic reasoning again with the same activity result. This means that the action is part of A1, however, more than one sub-activities share the same actions. Similarly, other events are added to AT1 $= \{e_1, e_2, e_3, e_4\}$ as they occurred with new *MakeTea* activity name which is a descendant class of *MakeDrink* and *KitchenADL*. Until now, only first stage of decision process is performed due to generic nature of the ADL actions. The next sensor observation, e_5, is attached to *Ginger* running any personalised actions. The activity name, *MakeTea* of A1 and the new sensor observation *Ginger*(e_5) is used to perform subsumption reasoning in the first stage of decision engine and returned inconsistency in ADL description error. In the second phase, the decision engine checks if the *Ginger*(e_5) sensor is part of an inhabitant's preference(s) stored in the triplestore and add it to A1. In this case, *spicyTea* preference was identified and as there were no sub-activity preference threads already active for A1, new thread Pref_1 was created along with other missing *spicyTea* actions.

$$\begin{aligned} \text{AT1} &= \{tbox\{name : makeTea,\ \text{s}:\ \text{e}_1,\ \text{e}_2,\ \text{e}_3,\ \text{e}_4\}\}, \\ &\quad \&\ \text{abox}[Pref_1[name : spicyTea, s : \{\text{e}_5\}\}, missing : \{\ldots\}]]\}. \end{aligned} \tag{6.7}$$

6.2.3 Semantic Segmentation Algorithm

The algorithm in Table 6.1 illustrates the segmentation process, use of decision engine (DE) and multithreading mechanism discussed in Sect. 6.2 to separate sensor

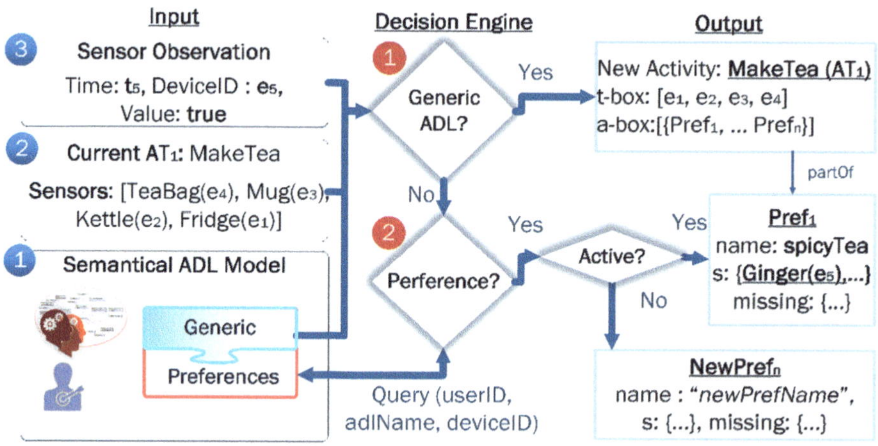

Fig. 6.4 Example of semantic-based decision engine with input and output data

observations. The algorithm is performed by the ADL threads manager and it is broken down into three stages. The first stage is to iterate over all the active $\mathcal{T}$-box threads (AT_n) and use the current list of sensor observations in each thread along with the observed sensor event (e_m) being investigated to refine a ADL inferencing result or assume start of new ADL. For simplicity, Algorithm Table 6.1 shows only the first iteration AT_1 is conducted.

The line 1 checks if there is $\neg\ni e_m$ in AT1 then perform $\mathcal{T}$-box and $\mathcal{A}$-box reasoning in stage two and three. Otherwise, e_m is assumed to be start of new ADL activity. Hence, new AT_{n+1} is created with e_m in line 12. The $\mathcal{T}$-box reasoning task in line 2 is performed by calling DE by taking three inputs: e_m, set of current sensor events in AT1 and $T = \{EC, SR, ET\}$ in KB. The new deduced ADL result (*Class c*) is evaluated for conflicts and if $c \sqsubseteq currentAT_1$ class then AT_1 is updated with c along with e_m; see lines 3 and 9. In the second stage, inhabitant's preferences are checked when conflicts in result is detected. All the $\mathcal{A}$-box threads are checked if e_m is part of active preference thread then add the event to $AboxT_a$ thread. Otherwise, any inhabitant (*userID*) preferences ($AboxT_a$) of a given ADL class c inferred for AT_1 is queried from the triplestore database (TDB) and new $\mathcal{A}$-box threads are created if matched; see lines 4–7. The final stage is where all the housekeeping for the sub-threads and the process of re-evaluating the session timeout window size and timeout cases based on the data of the segmented set of observations. Details of the semantic segmentation mechanism can be found in work [18, 19].

Table 6.1 The pseudocode of the semantic segmentation algorithm

Algorithm 1: Pseudocode for Semantical Segmentation Algorithm

Input: $e_m, T = EC, SR, ET, AT_1$

Output: void

```
if ¬∃e_m : AT_1(e_m) then
    Class c = DE.runTbox(e_m, AT_1, T)        /* 1) T-box reasoning */
    if ¬c ⊇ AT_1 then
        if ∃AT_1.AboxT_a(e_m) then
            AT_1.AboxT_a.add(e_m)             /* 2) A-box reasoning */
        else if ∃DE.queryTDB(e_m, AT_1.name, userID) then
            AT_1.addAboxT(e_m)                /* 2.1) create A-box thread */
    else
        AT_1 ≡ c(e_m)                         /* 1.1) update ADL classification */
    end
else
    AT_{n+1}(e_m)                             /* 1.2) create T-box thread */
end
closure(AT_1)                  /* 3) completion and timeout procedures */
```

6.3 Semantic Segmentation Lifecycle

This semantic segmentation approach has been implemented in a SOA based system. Key ontology modelling knowledge, multithreading process and reasoning tools has been highlighted in this section. Other technical details of the system are provided in Chap. 9 and Sect. 9.3.

6.3.1 *Ontological Modelling*

The generic knowledge for segmentation is represented using semantic web framework. This framework provides web ontology language OWL to formally express the complex knowledge into classes, relationships (object & data properties) and data (individuals) [20]. In addition, common vocabularies are used to represent the KB and encourage sharing across applications to create an ever-growing, human and machine-readable web of knowledge. There are a number of automatic reasoning tools available to read this KB to identify inexplicit facts based on relationship definition and the selection of a reasoner is elaborated in Sect. 6.3.3. The main goal of the ontological model is to express what, where and how the actions are required in order to satisfy a given ADL. For this, *EC, SR,* and *Pref* are modelled in three phases using ontology editor tool named Protégé [21]. Initially, *EC* concepts such as everyday objects (Obj_x), person (X_n), sensor characteristics (S_o) and location (L_m) were modelled as classes. Figure 6.5 illustrates the fragments of *EC* classes and their subclasses.

In the second phase, the *EC* classes are used to define SR between ADL classes and describe their actions iteratively using object properties. Figure 6.6 partially describes the *MakeTea* ADL in Protégé. The *MakeTea* ADL class inherit the properties described from super-classes and uses "*rdfs:subclassOf*" object property to define actions or the context to carry out the activity. The actions properties and the classes of everyday objects for the *MakeTea* ADL are described using object properties *hasAdding, hasContainer, hasHeatingAppliances, hasHotMealMaterial* and so on. These object properties can have characteristics and relationships between everyday objects classes and the ADLs. For instance, *hasHotDrinkType* object property has a *domain* of *MakeHotDrink* ADL class and *HotDrinkType* material as *range* property. This means that any everyday object that is a subclass of *HotDrinkType* is part of the actions defined for *MakeHotDrink* ADL class or its subclasses. These object properties are used to add further restrictions such as universal and existential quantification ($\forall$, $\exists$) using some and only, logical operations such as not, and, or ($\neg$, $\wedge$, $\vee$), and cardinality restrictions ($\leq$, $\geq$, $\equiv$). Other common operators are also available and can be used to increase the expressivity of the ADL model in terms of class, relationships and data. Similarly, the other 12 subclasses of *MakeDrink* and *MakeMeal* ADL classes are also described with relevant relationships. As multiple relationships with ADLs and everyday objects are created, the observed data (defined as individuals) with a set of assertation statements containing everyday object and object properties are used by the reasoning engine to automatically infer the type of the ADL class the actions in the individual belongs to.

Finally, the inhabitant specific preferences ($\mathcal{A}$-Box) are captured by creating individuals with a direct relationship with instances of sensors in order to avoid the inconsistency in ontology description for generic knowledge. In the generic knowledge, not all adding (ingredient) for *MakeTea* ADL are defined and ingredients such as *FreshGinger* and *CinnamonSticks* are subjective to the individual. Hence, forcefully adding ingredients in an instance that is the type of *MakeTea* ADL will result

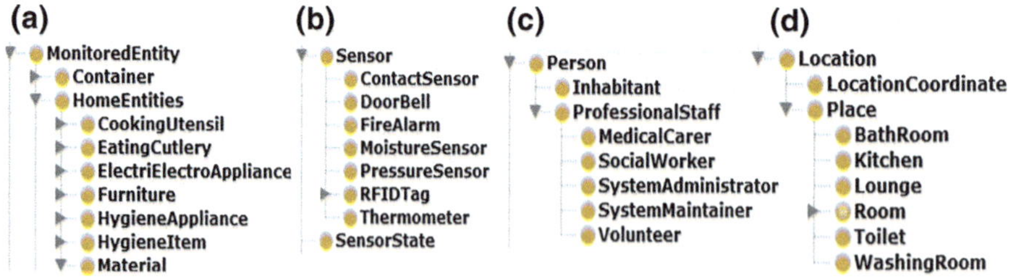

Fig. 6.5 Conceptualising environmental context (EC) with Protégé

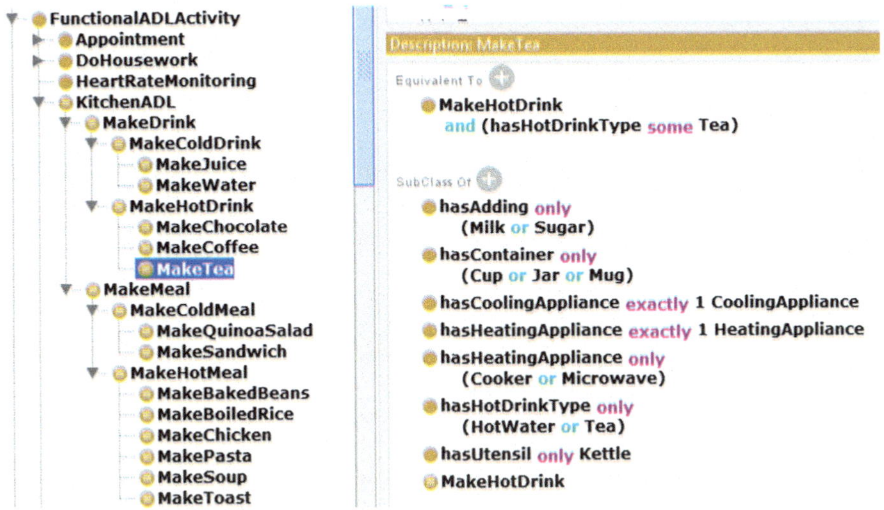

Fig. 6.6 An excerpt of *MakeTea* ADL with semantic relationship (SR) between EC in Protégé

in the inconsistent ontology as highlighted by the explanation window in Fig. 6.7. Therefore, instances of preferences are associated with the inhabitant and to a given ADL class which has a list of sensors that are attached to the everyday objects and other attributes. Figure 6.8 presents an example of three inhabitant preferences. The top section presents individual named, *Patient1_Preferences_IndianTea*, which has a type of *Preference* class for *MakeTea* ADL class along with a list of sensors using *hasSensor* object properties and data properties to describe other attributes such as preference name and creation timestamp. Similarly, other preferences are shown in the middle and bottom of the Fig. 6.8 to describe *MakeToast* and *MakeBakedBeans* preference.

Another method is available to layer the inhabitant specific and generic ADL ontology descriptions along with SWRL rules. This can be achieved by using the OWL API and Jena API to create and manipulate the model once generic and inhabitant specific models are combined, and rules are loaded into the memory. The reasoning can be performed using the Pellet reasoner and JESS rule engine after combining the generic and inhabitant specific ontology that is managed dynamically. However, the main limitation of this method is that the changes made to the inhabitant specific ontologies will need to be tracked along with the mechanism to resolve any conflicts in the knowledge that may arise. In addition, inhabitant specific reasoner will need to be created and maintained [22] at run-time. Hence, the amount of in-memory space, number of processing cores and computation power required can grow exponentially. This can potentially create high latency in segmenting individual sensor events and undermine the scalability of the approach. Therefore, the first method is selected as it is lightweight, and no inhabitant specific reasoner is required to be running. The SPARQL Inferencing Notation (SPIN) [23] rules or just a SPARQL query language

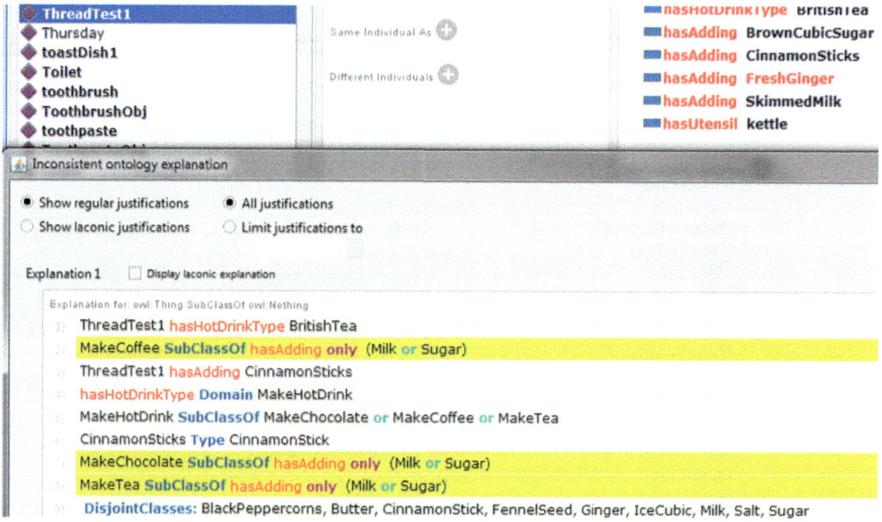

Fig. 6.7 Inconsistency on *hasAdding* object property due to the restrictions in *MakeTea* ADL class

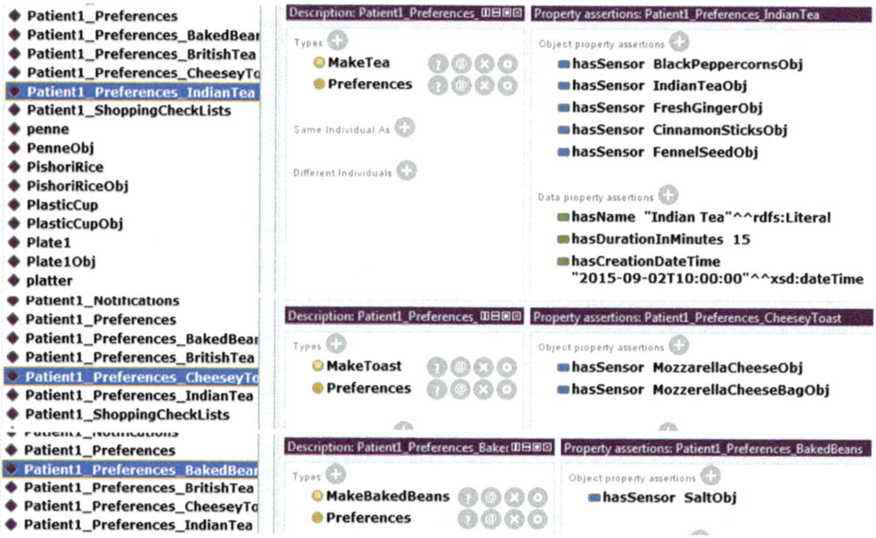

Fig. 6.8 Inhabitant preferences as individuals with a list of sensors

can be executed on the triplestore to retrieve multiple inhabitant's preferences for a given ADL class simultaneously. Therefore, this method is considered appropriate during the segmentation phase as the inhabitant's preferences can be scalable and has lower latency in terms of query time and there are no additional overheads for running multiple reasoners per inhabitant.

6.3.2 Multithread Segmentation Process

The multithread segmentation process is depicted in Fig. 6.9 where actions for *MakeTea* and *MakeToast* ADLs are performed concurrently. The generic and preferred actions are observed at a given time (t_n). The $\mathcal{T}$-box activity thread (AT1) is initially created when the *cupObj* sensor is activated at t_1. The AT1 continuously stores the events into the thread if the decision engine infers an association with generic ADL class in the ontological model or personalised preference(s). The object attached to the *cupObj* sensor is queried from the triplestore, added to new individual and incremental $\mathcal{T}$-box reasoning is conducted. The $\mathcal{T}$-box reasoning result indicates that the object is related to *ADLActivity* class with no conflicts with the model, hence the $\mathcal{A}$-box reasoning is not required to be executed. Next, the sensor event at t_2 is received and AT1 performs $\mathcal{T}$-box reasoning with observed sensor *fridgeObj* along with previous sensor(s), in this case, *cupObj*. The decision engine returned a new result, *KitchenADL* class and it was compared against the current *ADLActivity* class for equivalent or subsuming class. In this case, the subsuming condition is satisfied and stores the *cupObj* and *fridgeObj* sensor events in the AT1.

Similarly, *milkObj*, *kettleObj* and *indianTeaObj* sensor events are processed by AT1 where the ADL classes are incrementally classified, and the sensor events are stored in the thread. Since, the *freshGingerObj* sensor event is not described as part of a set of adding in the generic *MakeTea* ADL description, the decision engine returns with traceable conflicts. The decision engine then performs $\mathcal{A}$-box reasoning to find any inhabitant's preferences related to *MakeTea* ADL containing *freshGingerObj*. Multiple preferences could be returned, in this case, only one preference named, *Patient1_Pref_IndianTea* (P_1) is returned as a result of SPARQL query. A single $\mathcal{A}$-box sub-thread (ABT1) is created with other missing sensors and other relevant information from the preference into the thread. The ABT1 thread then inspects the incoming sensor events and updates the missing and matched sensors list independently. AT1 thread and the sub-thread(s) for $\mathcal{A}$-box reasoning can continue inspecting unfolding events in the data stream until the completion criteria are satisfied i.e. having no child ADL class and missing sensors in $\mathcal{A}$-box threads or a dynamic timeout mechanism for the ADL. The completion/timeout criteria for the ADL will be inspected in future work.

The next set of actions for *MakeToast* ADL are observed between t_8-t_{14} and inspected by AT1 but only one shared *fridgeObj* event is stored. The ADL manager running in parallel inspects the sensor events in the queue and detects *toastObj* is not part of the *MakeTea* ADL class in AT1 and ABT1 threads. Therefore, another

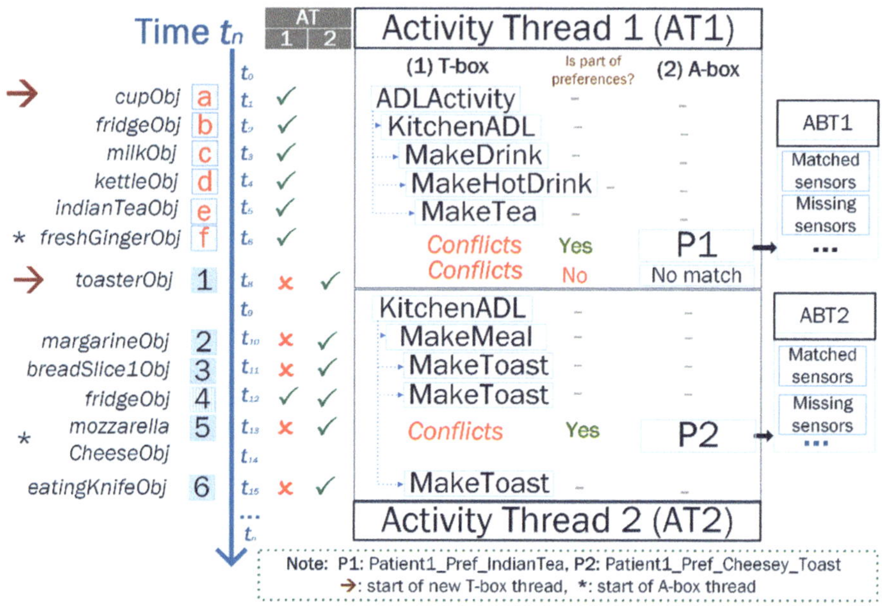

Fig. 6.9 Semantical segmentation process with concurrent actions for *MakeTea* and *MakeToast* ADL

$\mathcal{T}$-box activity thread (AT2) is created *MakeToast* ADL as depicted at the bottom-right of Fig. 6.9. The same process is described for AT1 is executed for the AT2 thread to capture events from t_{10}-t_{15} to AT2 thread with one conflicting *mozzarellaCheeseObj* observation. Therefore, the ABT2 thread is created when identified by decision engine that *mozzarellaCheeseObj* is part *Patient1_Pref_CheeseyToast* (P2) to perform the *MakeToast* activity.

6.3.3 *Reasoner and Supporting Tools*

A reasoner is a software tool developed to perform $\mathcal{A}$-box and $\mathcal{T}$-box reasoning by the decision engine to perform tasks such as consistency check of the ontological model and derive new facts from the KB dataset. There are a number of reasoners developed over the years and most of them support first-order predicate logic [15] reasoning or procedural reasoning (perform forward and backward chaining). Some of the key requirements for selecting a reasoner are that it supports the incremental classification for only the part of ontology that was affected by the changes [24], full description logics (DLs) family support for higher expressivity, rules support, justification of conflicts, low latency in classification and support both $\mathcal{T}$-Box and $\mathcal{A}$-Box reasoning. Studies in [15, 16] describe a number of popular reasoners using large ontologies, compare against their key features and categorise according to their characteristics. The incremental Pellet reasoner has been selected as it supports most requirements stated above along with being open source and supported by a number of application programming interfaces (APIs) and ontology editors such as Protégé and NeOn toolkit. OWL API and Jena API both support the Pellet reasoner to programmatically perform reasoning, querying and KB manipulation. Jena API further supports other reasoners to be Integrated easily. Although, the pellet reasoner takes up higher heap space and has higher delay time than FaCT + when performing concept satisfiability checking after classification but outperforms in subsumption query.

6.4 An Example Case Study

6.4.1 Experiment Design

The actions for three ADLs are scripted in no particular order to perform with only generic actions and another with the inhabitant's preferences; namely, *MakeTea, MakeToast* and *MakeBakedBeans*. The relevant actions for the generic(G) ADL and some inhabitant's preferences (P) are described in Table 6.2. These three ADLs are first tested individually in random order and then combined to create composite activity scenario; incremental, concurrent and parallel; see Table 6.3. A total of 30 activity scenarios (6 for single and 24 for composite ADLs for both G, and G + P actions) were created for the experiment and a thread simulated each scenario with sensor events occurring at 10 ms interval. The sensor events contained a timestamp, name, sensor type, and binary data. The degree of accuracy to recognise an activity scenario is calculated in percentage by matching and tallying actual sensors events segmented correctly and it divided by the total number of sensors events activated for each ADL. The average classification time is calculated by taking sensor observation segmented time by the reasoner minus the sensor observation time recorded for each activity scenario. The unexpected sensor observations within the activity scenario are omitted and recorded separately when calculating the accuracy and average classification time for the activity. In addition, a number of duplicate activity threads created in the activity scenario are also recorded to see the effect on the overall classification times. The Samsung S6 edge smartphone running 6.0.1 Android OS was used and the web service was deployed on the HP EliteBook Folio 1040 G2 with the i7 2.60 GHz processor, 2 cores, 4 logical processors and 8 GB RAM. The binary sensor events are currently simulated due to a limited number of sensors and time.

Table 6.2 Examples of sequential actions of single activities

Activity	Type	Related actions/sensors attached to objects	#
Make tea	G	KettleObj, Cup1Obj, TeaJarObj, IndianTeaObj, KitchenSinkTap1Obj, SugarJarObj, FridgeObj, Milk1Obj, Spoon2Obj	9
	P	[FreshGingerObj], [CinnamonSticksObj], [BlackPeppercornsObj], [FennelSeedObj]	4
Make baked beans	G	Spoon1Obj, HenzBeansCan1Obj, HenzBeansObj, CanOpener1Obj, MicrowaveBowl1Obj, MicrowaveObj, Plate1Obj, EatingKnifeObj	8
	P	[SaltObj]	1
Make toast	G	Plate1Obj, BreadPacket1Obj, BreadSlice1Obj, ToasterObj, FridgeObj, MargarineObj, EatingKnifeObj	7
	P	[MozzerellaCheeseBagObj], [MozzarellaCheeseObj]	2

Note Generic (G)/Preference (P) actions, [SensorName]—User preference item, #—number of sensors

Table 6.3 Combinations of simple activities

Activity comb.	ADL sequences	Expected no. threads	Actions	
			Gen. (G)	+pref. (G + P)
AC1	MakeTea, MakeToast	2	16	22
AC2	MakeTea, MakeBakedBeans	2	17	22
AC3	MakeToast, MakeBakedBeans	2	15	18
AC4	MakeToast, MakeBakedBeans, MakeTea	3	24	31
AC5	MakeBakedBeans, MakeTea, MakeToast	3	24	31
AC6	MakeTea, MakeToast, MakeBakedBeans	3	24	31
Total		15	120	155

6.4.2 Results and Discussions

The average segmentation time taken per sensor event for single activity is 3971 ms in contrast to 62183 ms for composite ADL scenarios as shown in Tables 6.4 and 6.5. The result in Table 6.4 shows that all the sensor events for a single activity case scenario were adequately placed in the correct thread with 100% accuracy. Only the *MakeTea* activity case scenario created additional threads with more than double the average time when processing 9 generic actions and 4 preferred actions. On

Table 6.4 Single activity performed in no specific order with generic and personal preferences

Activity	Type	In relevant thread	Unexp. actions in thread(s)*	Excess thread (s)	Avg. time (ms) +
MakeTea	G	9	0	0	2394.67
MakeToast	G	7	0	0	2468.57
MakeBaked Beans	G	8	0	0	2372.25
MakeTea	G + P	13	0	1	10828.85
MakeToast	G + P	9	0	0	3786.87
MakeBaked Beans	G + P	9	0	0	1972.44
Total	6	55/55	0	1	3970.61 (avg)

Note * excludes additional thread(s) actions, + including excess threads

the other hand, Table 6.5 shows 20 out of 24 activities performed in a composite manner or 572 out of 585 sensor events were added to the relevant thread, giving 97.8% accuracy. However, the segmented activity threads captured a total of 71 additional unexpected sensor events in the segmented threads which are not necessarily incorrect, i.e., multiple spoon objects or heating/cooling appliances when performing multiple activities interweavingly. Furthermore, 29 additional threads were created and failed to classify any ongoing activity.

Although, previous studies use varying ADL models, datasets, sensors and platforms, use scenarios, and etc., the key features and final outcomes for the recent KD studies presented in Sect. 6.1.1 is discussed instead. The accuracy of single and composite activity segmentation for evidential theory-based approach [14] is 81.8% and 76.2% on average and ontology and temporal [10] achieved 100% and 88.3%, respectively. Therefore, there is a significant evidence that the proposed approach improves the accuracy of sensor segmentation with 100% and 97.8%, respectively. In addition, user-preferences are taken into consideration by adopting basic query-based approach and automatic Pellet reasoner for generic KB reasoning compared to their counterparts which adapt solely query-based approach inheriting classical multi-query optimisation problem in [8, 12]. Nevertheless, one of the benefits for adapting multi-query approach is that higher performance and scalability can be achieved, however, suffer from the expressivity capabilities of KB due to explicit query development/maintenance efforts and the ability to use automatic reasoners.

The proposed method in this chapter seeks to strike a balance between automation by taking advantage of expressive ontology with incremental Pellet reasoning feature and performance of query-based approach to manage the changing user-preferences. The average segmentation time information is not available in the presented KB studies; however, the proposed approaches observes 3971 ms and 62183 ms with sensors events activated at the 10s interval for simple and composite activities scenarios. These results are still not suitable for the real-time system at this stage. However, the optimisation opportunities such as multi-thread safe reasoning [25], ADL threads management, parallel programming, partitioning workload to graphics processing units (GPUs) [26] Scaling parallel rule-based reasoning [26] and using a machine with higher number of cores (i.e., quad-core, octa-core CPU or higher) to support more concurrent or parallel threads execution at same time remain an open challenge. Table 6.6 presents a summary of the key components of the recent KB studies presented in Sect. 6.1.1 against the proposed semantic segmentation approach in this chapter.

6.5 Summary

The semantic-based segmentation approach combines generic knowledge conceptualised as an ontological model and inhabitant specific preferences to conduct a specific ADL as asserted individual. Upon sensor activation, the event is inspected by one or more active ADL threads running in parallel. Each ADL thread relies on a two-stage decision engine to find any association with observed sensor event. The decision engine conducts $\mathcal{T}$-box reasoning with generic KB in the first stage and $\mathcal{A}$-box reasoning with observed sensor event and inhabitant specific preferences by querying the triplestore in the second stage. The second stage of decision engine is only invoked when the use of entity on which observed sensor is attached to has a contradiction or not been explicitly specified in generic ADL description. The ADL

Table 6.5 Multiple activities performed in a composite manner

	Activity Comb.	Type	All actions in threads?		Excess thread(s)	Unexpected actions in the threads*	Total avg. time (ms)
Inc.	AC1	G	✔	16	1	1	36330.64
	AC2	G	✔	17	1	4	41543.17
	AC3	G	✔	15	1	1	30354.98
	AC4	G	×	15/24	3	3	95819.25
	AC5	G	✔	24	1	5	60742.14
	AC6	G	✔	24	1	6	72690.97
	AC1	G + P	✔	22	1	1	54949.21
	AC2	G + P	✔	22	0	5	21905.05
	AC3	G + P	✔	18	0	1	12561.28
	AC4	G + P	×	31	3	3	99807.19

(continued)

Table 6.5 (continued)

	Activity Comb.	Type	All actions in threads?		Excess thread(s)	Unexpected actions in the threads*	Total avg. time (ms)
	AC5	G + P	×	30/31	1	4	62016.20
	AC6	G + P	✔	31	1	3	87298.32
Con.	AC1	G + P	✔	22	1	0	56752.83
	AC2	G + P	✔	22	1	5	23993.51
	AC3	G + P	✔	18	2	1	64074.61
	AC4	G + P	✔	31	1	1	70289.79
	AC5	G + P	✔	31	2	6	131784.92
	AC6	G + P	✔	31	2	5	181894.97
Par.	AC1	G + P	×	21/22	2	0	43055.55
	AC2	G + P	✔	22	0	3	8309.10
	AC3	G + P	×	16/18	1	0	35944.94
	AC4	G + P	✔	31	1	4	63737.04
	AC5	G + P	✔	31	1	5	77355.87
	AC6	G + P	✔	31	1	4	59173.90
	Total	24	572/585		29	71	62182.73 (avg.)

Note * excludes additional thread(s) actions, ++ including excess threads

thread discards the observed event when decision engine has failed to find any relationship. When the whole set of active ADL threads fail to find any relevance for a given sensor event, a new ADL thread is created. The approach leverages between the incremental Pellet reasoner, OWL & Jena API, and the notion of multithreading. The proposed method was implemented and tested against 30 test scenarios. The results indicate an improvement in segmentation accuracy compared to the counterpart studies with 100% and 88.3% for single and composite ADL scenarios with an average time of 3971 ms and 62183 ms. The main bottlenecks for high processing

Table 6.6 Summary of recent KB approaches

Studies/features	C-SPARQL [8], 2010	Evidential theory [14], 2013	Onto. and temporal [9], 2014	AALISABETH [12], 2015	Proposed
Knowledge expressivity	High	High	High	High	High
SPARQL query support	Yes	Yes	Yes	Yes	Yes
Automatic reasoner support	No	No	Yes	No	Yes
Direct stream inspection	No	Yes	Yes	No	Yes
RDF stored	Yes	NA	Yes	No	Yes
User prefs. support	No	No	No	No	Yes
Sliding window support	Yes (Fixed size)	No	Yes	Yes	No (Future work)
Potential scalability issue	Low	Med.—High	Med.	Low	Med.—High
Accuracy: S; C (%)		81.8; 76.2	100; 88.3	–	100; 97.8
Average time: S; C (ms)	–	–	–	–	3971; 62183

time are the synchronised incremental reasoning and duplicate ADL threads creation which ultimately created additional reasoning tasks and slowed down the overall process on the machine which was limited to two cores. This could be addressed by adapting Fork/Join parallelism framework [27] to efficiently split and manage tasks over multiple cores machine and utilise graphical processing unit (GPU) to increase performance.

References

1. Faria DR, Vieira M, Premebida C, Nunes U (2015) Probabilistic human daily activity recognition towards robot-assisted living. 2015 24th IEEE international symposium on robot and human interactive communication (RO-MAN), pp 582–587
2. Zhong Y (2017) A theory of semantic information. China Commun 14:1–17
3. Tarski A (1944) The semantic conception of truth: and the foundations of semantics. http://www.jstor.org/stable/2102968?origin=crossref
4. Vickers P (2013) Understanding visualisation: a formal foundation using category theory and semiotics. IEEE Trans Vis Comput Graph X, 1–14

5. Wang Y (2017) Formal rules for concept and semantics manipulations in cognitive linguistics and machine learning. In: 2017 IEEE 16th international conference on cognitive informatics cognitive computing (ICCI*CC), pp 43–50
6. Rafferty J, Nugent CD, Liu J, Chen L (2016) From activity recognition to intention recognition for assisted living within smart homes. IEEE Trans Human-Mach Syst 1–12
7. Meditskos G, Dasiopoulou S, Kompatsiaris I (2015) MetaQ: a knowledge-driven framework for context-aware activity recognition combining SPARQL and OWL 2 activity patterns. Pervasive Mob Comput
8. Della Valle E, Grossniklaus M (2010) C-SPARQL: a continuous query language for RDF data streams. Int J Semant Comput 04:3–25
9. Okeyo G, Chen L, Wang H, Sterritt R (2012) A hybrid ontological and temporal approach for composite activity modelling. In: Proceedings - 12th IEEE international conference on trust, security and privacy in computing, trustcom-2012 - 11th IEEE int. conference on ubiquitous computing *and* communications, IUCC-2012, pp 1763–1770
10. Okeyo G, Chen L, Wang H (2014) Combining ontological and temporal formalisms for composite activity modelling and recognition in smart homes. Futur Gener Comput Syst 39:29–43
11. Skillen KL, Chen L, Nugent CD, Donnelly MP, Burns W, Solheim I (2014) Ontological user modelling and semantic rule-based reasoning for personalisation of help-on-demand services in pervasive environments. Futur Gener Comput Syst 34:97–109
12. Culmone R, Giuliodori P, Quadrini M (2015) Human activity recognition using a semantic ontology-based framework. Int J Adv Intell Syst 8:159–168
13. Naeem U (2015) Activities of daily life recognition using process representation modelling to support intention analysis. Int J Pervasive Comput Commun 11:347
14. Hong X, Nugent CD (2013) Segmenting sensor data for activity monitoring in smart environments. Pers Ubiquitous Comput 17:545–559
15. Abburu S (2012) A survey on ontology reasoners and comparison. Int J Comput Appl 57:33–39
16. Dentler K, Cornet R, Ten Teije A, De Keizer N (2011) Comparison of reasoners for large ontologies in the OWL 2 EL profile. Semant Web 2:71–87
17. De Giacomo G, Lenzerini M (1996) TBox and ABox reasoning in expressive description logics. In: Proceedings of fifth international conference on the principles of knowledge representation and reasoning, pp 316–327 (1996)
18. Triboan D, Chen L, Chen F, Wang Z (2017) Semantic segmentation of real-time sensor data stream for complex activity recognition. Pers, Ubiquitous Comput
19. Triboan D, Chen L, Chen F, Fallmann S, Psychoula I (2017) Real-time sensor observation segmentation for complex activity recognition within smart environments. In: 2017 IEEE 14th international conference on ubiquitous intelligence and computing (UIC 2017), San Francisco
20. Riboni D, Bettini C (2011) OWL 2 modeling and reasoning with complex human activities. Pervasive Mob Comput 7(3):379–395
21. Stanford University, University, S Protégé
22. Volz R, Staab S, Motik B (2003) Incremental maintenance of materialized ontologies. Lect Notes Comput Sci 2888(2003):707–724
23. W3C: SPIN - overview and motivation. https://www.w3.org/Submission/spin-overview/
24. Cuenca Grau B, Halaschek-Wiener C, Kazakov Y (2007) History matters: incremental ontology reasoning using modules. Lecture notes in computer science (including subseries Lecture notes in artificial intelligence (LNAI) and lecture notes in bioinformatics). LNCS, vol 4825, pp 183–196
25. Ren Y, Pan JZ, Guclu I, Kollingbaum M (2016) A combined approach to incremental reasoning for EL ontologies. Lecture notes in computer science (including subseries Lecture notes in artificial intelligence (LNAI) and lecture notes in bioinformatics). LNCS, vol 9898, pp 167–183
26. Peters M, Brink C, Sachweh S, Zündorf A (2014) Scaling parallel rule-based reasoning. Lecture notes in computer science (including subseries Lecture notes in artificial intelligence (LNAI) and lecture notes in bioinformatics). LNCS, vol 8465, pp 270–285
27. Ponge J Fork and join: java can excel at painless parallel programming too!. http://www.oracle.com/technetwork/articles/java/fork-join-422606.html

Chapter 7
Composite Activity Recognition

7.1 Introduction

Inhabitants within smart homes (SH) typically perform Activities of Daily Living (ADLs) in complex patterns. For instance, an inhabitant may perform two (or more) activities in sequence or in parallel. Whenever activities are performed sequentially or in parallel, there will be underlying inter-activity dependencies among the activities involved. These inter-activity dependencies should be encoded during activity modelling so as to support activity recognition in the presence of complex activity patterns, e.g. composite activities. Applications that provide SH inhabitants with services, e.g. assistive services, should be able to correctly identify both simple and composite activities. Activity recognition is the process of tracking users and identifying the activities they are performing. It involves activity sensing, activity modelling, and activity inference. Activity sensing is responsible for monitoring users and their situated environment to obtain sensor data streams. Activity modelling creates computational activity models that are used to analyse and classify collections of sensor data into activities. Activity inference uses relevant algorithms to process sensor data against computational activity models to identify the ongoing activity.

In this chapter we categorise activities as actions, simple activities, and composite activities. An action is an atomic (or indivisible) activity, e.g. grasping the fridge door. A simple activity is an ordered sequence of actions, e.g. preparing coffee. Finally, a composite activity is a collection of two or more simple activities occurring within a given time interval, e.g. preparing dinner and washing dishes. Composite activities can be further categorised as sequential or multi-task activities. A sequential activity is a sequence of activities that occur in consecutive time intervals, i.e., there is temporal dependency between constituent activities. A multi-task activity occurs when a single user performs two or more activities simultaneously or when multiple residents occupy a smart environment and perform activities concurrently.

Activity recognition has been widely investigated using three categories of approaches, namely, data-driven (DD), knowledge-driven (KD), and hybrid activity recognition approaches as discussed in previous chapters. In data-driven activ-

L. Chen and C. D. Nugent, *Human Activity Recognition and Behaviour Analysis*, https://doi.org/10.1007/978-3-030-19408-6_7

ity recognition, activity models are learnt from pre-existing datasets using existing well-developed machine learning techniques. Activity inference is then performed against the learnt activity models whenever sensor data is obtained. In knowledge-driven activity recognition, knowledge engineers and domain experts specify activity models using a knowledge engineering process. The activity models capture common-sense and domain knowledge about activities. Artificial intelligence-based reasoning techniques are then used to infer activities from the models whenever sensor data is obtained. Hybrid activity recognition approaches combine data-driven and knowledge-driven techniques.

Simple activity recognition has been widely explored in DD, KD, and hybrid activity recognition. However, composite activity recognition is only investigated to a limited extent in DD [1–6] and hybrid [7–9] activity recognition communities. In the KD activity recognition research community, the recognition of composite activities still remains largely unexplored. This challenge can be attributed to the two tasks of activity modelling and activity inference. Composite activity modelling is a challenge because activity models must capture and reason with inter-activity dependencies that are typically encoded as temporal knowledge [10]. Moreover, mechanisms are needed to process sensor data against the resulting composite activity models to infer the ongoing activities [11].

The use of ontologies in activity modelling and activity recognition has spurred interest but the focus has largely been on simple activities [11–13]. Ontological activity modelling can be used to define activity ontologies that describe activities and their characteristics [12, 13]. The resulting activity ontologies represent activity models for mostly simple activities and support semantic reasoning for activity recognition. To support composite activity modelling and recognition, we have developed a novel activity modelling approach that combines ontologies and temporal knowledge to create activity models that represent inter-activity dependencies using temporal relationships. The approach enhances ontological activity models by adding qualitative temporal knowledge based on Allen's temporal logic relations [14]. It is worth pointing out that the study presented in this chapter is contextualised in a single-resident SH environment within which the user performs both simple and composite activities.

In this chapter we introduce a novel hybrid approach to composite activity modelling and recognition. The combination of ontological and temporal knowledge representation formalisms provides a more expressive representation formalism required for representing and modelling the complex ontological and temporal relationships of composite activities. We also develop generic activity models for composite activities based on the presented approach. These include three core elements, namely ontological activity models, temporal activity models and entailment rules; each element models a specific aspect of composite activities. The generic models can be applied to modelling composite activities in different application scenarios. In addition, we create reusable activity models for ADLs in the context of smart homes for the purpose of illustration, testing and evaluation. Finally, we develop an integrated system architecture for composite activity recognition and a unified activity recognition algorithm. The algorithm can reason over sensor data streams against composite activity models to perform real-time progressive activity recognition for both simple and composite activities.

7.2 Related Work

In the DD activity recognition community, existing approaches capable of both simple and composite activity modelling and recognition include hidden Markov models (HMM) [2], interleaved HMM [1], factorial conditional random fields (FCRF) [4], skip-chain conditional random fields (SCCRF) [3, 5, 6] and mining of emerging patterns [15]. DD approaches have the ability to handle uncertain knowledge and are based on well-explored machine learning based techniques. They also have the advantage to handle temporal information that can capture short- and long-term temporal dependencies, e.g. inter-activity relationships and activity history, thereby making them suited to composite activity recognition. The main drawback is that large amounts of initial training data are needed to learn the activity models. As users perform activities in a variety of ways, all these activity variants must be present in the data set if they are to be successfully learnt, modelled and subsequently recognized. In most cases it is difficult to obtain representative and sufficient data sets to be used for learning activity models, thus leading to the "cold start" problem. In addition, users perform activities in different manners; as a result, models learnt from one user's datasets would not be reused by another user, which results in reusability problem.

KD activity recognition approaches use knowledge representation formalisms to provide explicit activity models which can be processed by artificial intelligence-based inference for activity recognition. The KD has a number of strengths. For instance, it is grounded in logic theory making it possible to capture the semantics of a domain and support automated reasoning. It also allows common sense domain knowledge and heuristics associated with activities to be incorporated into activity models. Domain knowledge is especially important in modelling complex real-world activity scenarios, e.g. interleaved and concurrent activities. Moreover, it can support reuse and knowledge sharing between applications. Domain knowledge and common-sense knowledge is essentially common across applications, hence the ability to encode and share it would make application development easier through reuse. The KD approach has been used for simple activity modelling and recognition, but little work has been done in composite activity recognition. Saguna et al. [16] addressed both simple and composite activity recognition by combining ontological and spatio-temporal modelling and reasoning. It uses the notion of context-driven activity theory (CDAT) to encode context information in order to model both primitive actions and simple activities. The resulting models are combined with ontological situation models and used to infer interleaved and concurrent activities. The authors derive models of situations, based on spatio-temporal information, from the context spaces theory [17], and use the resulting situations in activity inference. Essentially, it uses a layered approach to activity recognition consisting of atomic activity recognition (using machine learning techniques), simple activity recognition (using ontological inference), and finally interleaved and concurrent activity recognition (using rule-based inference).

Hybrid approaches to activity recognition combine techniques from data-driven and knowledge-driven activity recognition. So far, only Markov logic networks (MLN) [8] and HMMs with Allen logic [9] have been used to support activity recognition of simple activities, and interleaved and concurrent activities. Both approaches encode and use temporal knowledge but rely on automatically extracting the relevant temporal patterns from data sets. The main strength of the hybrid approaches is that they can model and recognize a range of simple and composite activities due to their ability to encode rich domain knowledge, e.g. temporal knowledge, and still utilize well-developed learning and probabilistic models. However, they suffer the "*cold start*" problem just like the DD approaches. Since our approach requires activity models to be specified based on domain knowledge, it overcomes the "*cold start*" problem.

Our work follows the KD approach but differs from Saguna's work, in two main aspects. First, the former requires the training of atomic activity recognition models from data sets. Secondly, it uses an ad hoc method to encode both temporal and spatial knowledge into activity ontologies. Basically, it simply captures temporal and spatial knowledge as properties of ontology concepts, a strategy that ignores reasoning and querying challenges arising from processing temporal or spatial knowledge in ontologies. Our work adopts a systematic and clear method for encoding and reasoning with temporal knowledge based on 4D-fluents [18] and therefore provides a clear mechanism for seamlessly integrating and exploiting qualitative temporal knowledge in activity recognition for both simple and composite activities. We believe this research enriches the literature and advances the research frontiers of the knowledge-driven approach to activity recognition.

7.3 A Hybrid Approach to Composite Activity Modelling

ADLs possess several unique characteristics that make activity modelling a difficult task. Firstly, there are over 20 categories of ADLs [19–21]. These include dental care, hygiene, bathing, dressing, using the toilet, drinking, transferring, mobility, orientation to time, driving/ using public transport, managing finances, drink preparation, use of the telephone, food preparation, housework, communication, shopping, eating, orientation to space, and games and hobbies [19, 22, 23]. Furthermore, each category can be classified to activities at multiple levels of granularity. For example, the activity Food Preparation can be broken down to child activities such as '*Prepare Coffee*', '*Prepare Toast*, etc. Also, '*Prepare Coffee* can further be classified into its child activities '*Prepare Espresso*, '*Prepare Latte*', etc. Thus, Food Preparation is coarse grained, whereas '*Prepare Coffee* is fine grained. Therefore, activity modelling approaches should support the different categories and granularities of activities.

Secondly, most ADLs involve performing a number of actions, with the ordering of actions dependent on an individual's preferences or abilities leading to a large number of ADL variants. Activity models should encode this activity diversity. Thirdly, the performance of activities may continuously change, e.g. activity duration or the sequence of objects can change, based on the users' abilities or preferences. Activity models, therefore, should be flexible to accommodate these variations. Fourthly, users also perform activities using complex patterns, such as interleaved and concurrent activities. Therefore, activity models should encode such complex relationships. Fifthly, activities are performed under different contexts, e.g. specific locations, objects, time, space, and goals. This is even more evident in composite activities. The resulting contextual information should be used to characterize activities. For instance, composite activities can be described by specifying inter-activity relationships using temporal or spatial information. In general, activities are characterized by rich temporal information, e.g. repetitive time patterns, temporal sequences, temporal duration, and time instants or intervals. For example, the occurrence of two activities A and B within the same time interval can represent a temporal inter-activity dependency that signals the occurrence of a composite activity.

The range of characteristics discussed above constitutes domain knowledge and heuristics upon which we have built a high-level conceptual activity model, as shown in Fig. 7.1, for composite activities based on the conceptual activity model that was proposed by Chen and Nugent [13]. The conceptual activity model in [24] describes an activity based on contextual elements (i.e., identity, time, space, actor, related activities, resources, environment elements, and goals) and properties that support inference (i.e., conditions and effects). However, its main limitation is that it does not explicitly provide a means to encode temporal inter-activity dependencies that typically characterize composite activities. This is because it ignored the important role that the model of change plays in composite activity modelling. Therefore, to support both simple and composite activity modelling, we have added two properties to the Time concept, i.e., temporal reference and model of change, to produce the revised model. The temporal reference is needed for both simple and composite activities, whereas the model of change is only mandatory for composite activity modelling. The temporal reference indicates the time interval or time instant that a given simple or composite activity occurs. The model of change represents the property that within a given temporal inference, a composite activity consists of two or more simple activities, whereby each simple activity can be identified by its respective temporal reference.

Ontological activity modelling has been used to create simple activity models as ADL ontologies [11, 25]. In this case, ADL activities are structured in a hierarchical tree with the most specific ADL descriptions represented as leaf concepts—all leaf concepts have no child classes. Each concept is associated with a number of role (property) restrictions. All child concepts inherit all the roles of their parent concepts but may specify further constraints. A generic activity refers to an ADL class that has associated descendant classes; whereas, a specific activity (the so-called leaf activity)

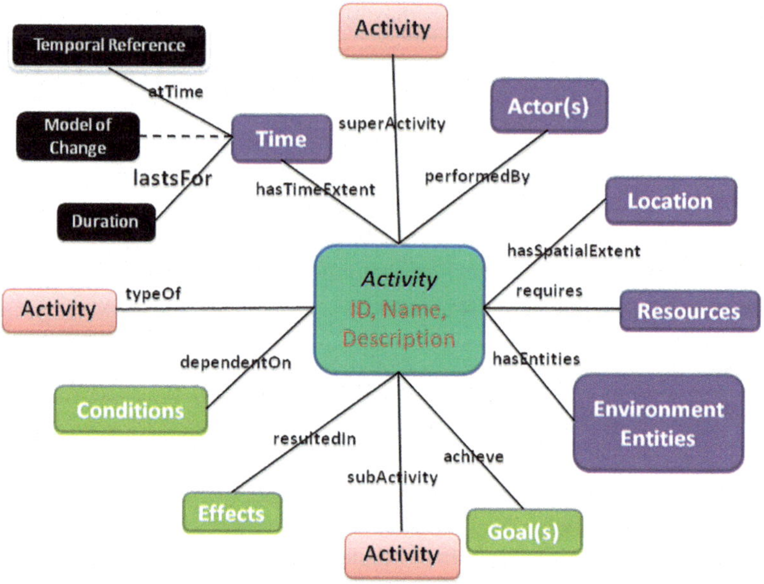

Fig. 7.1 The enhanced conceptual activity model

is an activity with no descendant classes in the ontology. For instance, '*Bathroom ADL*' is a generic activity, while its descendants '*Have Bath*' and '*Brush Teeth*' can be specific activities. Nevertheless, the aforementioned approach does not work for modelling composite activities. Despite capturing temporal information, e.g. time and duration, pure ontological modelling does not support temporal inference. For instance, OWL DL [26] only allows ontologies to capture temporal knowledge but does not support temporal reasoning and querying. To model composite activities, the approach to activity modelling has to be able to capture and model temporal inter-activity dependencies, and further support temporal reasoning. For this purpose, we have proposed the ontological and temporal approach to activity modelling, which are described in detail below.

7.3.1 Representing Temporal Knowledge in Ontologies

Description logics (DL) [27, 28] provide a mechanism that uses concepts, relations, and axioms for representing and reasoning with domain knowledge. Web ontology language (OWL) [26], a semantic web ontology language based on DL, provides a set of constructors and axioms for creating ontologies. In addition, it allows axioms for specifying subsumption, equivalence, disjointness, as well as property characteristics to be defined. The constructors, axioms, and DL equivalents are shown in Table 7.2.

The symbols used in DL formulas are $\mathcal{C}$ and $\mathcal{D}$ for concepts; r_i for role or property names; x_j for an instances; and n a non-negative integer.

On the other hand, temporal logic allows representation of and reasoning with temporal knowledge. Qualitative temporal knowledge naturally occurs in humans' activities, e.g. the user performs two activities, one after the other. Such qualitative temporal knowledge can be used to model complex relationships between activities that represent composite activities, e.g., sequential, and interleaved and concurrent relationships. In essence, each composite activity can be viewed as an activity that has changing relationships with other simple or composite activities. For instance, a composite sequential activity relates to two activities that occur in consecutive time intervals. In general, temporal knowledge allows knowledge at a particular moment of time and the notion of change in knowledge to be encoded and reasoned with. Therefore, a temporal representation specifies a temporal reference and model of change. The temporal reference captures order in the sequence of events using either point-based or interval-based time representation. The model of change captures the changing relationships between individuals relative to the temporal reference. The two aspects (i.e. change and temporal reference) can be used to capture complex relationships between activities, e.g., sequential, and interleaved and concurrent relationships. This can be achieved by using an appropriate temporal knowledge representation mechanism to encode qualitative temporal knowledge that naturally occurs in humans' activities, e.g. the user performs two activities, one after the other.

7.3.2 A Hybrid Ontological and Temporal Approach

Representing temporal knowledge in OWL is a challenge because OWL only supports unary and binary relations, while adding a temporal dimension requires at least a ternary relation. Therefore, we adopt the 4D-fluents approach [35, 18] to add a temporal model as a layer on top of the underlying DL. The 4D-fluents approach uses two fundamental building blocks, namely, time slices and fluents, to provide a vocabulary to represent dynamic temporal parts of individuals. It represents concepts that have a temporal extent as 4-dimensional objects, with the fourth dimension being the time, captured as time slices. The time slices represent the temporal parts of a specific entity at specific moments of time and the concept itself is then defined as an aggregate of all of its time slices. Time instances and time intervals are represented as instances of a time interval class. The instances are then associated with time slices to relate them with concepts varying in time. On the other hand, fluents are properties that hold at specific moments in time, whether interval or instant. In essence, the fluent property holds among two time slices. Changes occur on the properties of the temporal part of the ontology while keeping the entities of the other parts of the ontology unchanged. The 4D-fluents approach is chosen because it preserves OWL semantics when incorporating temporal knowledge into OWL ontologies and can therefore exploit existing OWL reasoning support. By combining ontological and temporal representation we can obtain a hybrid representation that not only encodes temporal knowledge but also supports inference with such knowledge.

The main idea that the hybrid approach uses for composite activity modelling is that within a time interval (a temporal reference), a composite activity can be characterised by one or more simple activities, and the simple activities involved can vary within sub-intervals of the main interval (model of change). We refer to models in which it is not mandatory to represent the model of change as static activity models, whereas those that encode both the temporal reference and change are denoted as dynamic activity models.

When the 4D-fluents approach is extended with qualitative relations [14], e.g. Allen temporal logic relations [14], it can model relations that are necessary for encoding composite activities. Allen's temporal logic refers to a constraint-based representation that uses a temporal interval as a primitive to support qualitative temporal knowledge representation and reasoning. It is based on the idea that much of the temporal knowledge is relative and so can be mapped into relations between intervals. The approach uses thirteen interval relations (shown in Table 7.1) that are considered adequate to express any relationship that can hold between two time intervals.

In this work we combine ontologies and temporal knowledge representation to create activity models for both simple and composite activities. To enable the resulting models to be exploited in composite activity recognition, interval relations and inference rules are used to provide procedural inference. In the next section, we apply this approach to generate activity models of simple and composite activities (Table 7.2).

Table 7.1 Thirteen interval relations

Relation	Symbol	Inverse symbol	Pictorial illustration
X before Y	<	>	XXX YYY
X equal Y	=	=	XXX YYY
X meets Y	m	mi	XXXYYY
X overlaps Y	o	oi	XXX YYYYY
X during Y	d	di	XXX YYYYYYY
X starts Y	s	si	XXX YYYYYYY
X finishes Y	f	fi	XXX YYYYY

Table 7.2 OWL constructors, axioms and dl syntax

OWL constructor	DL syntax	OWL axiom	DL syntax
IntersectionOf	$C \sqcap D$	SubClassOf	$C \sqsubseteq D$
UnionOf	$C \sqcup D$	EquivalentClass	$C \equiv D$
ComplementOf	$\neg C$	SubPropertyOf	$r1 \sqsubseteq r2$
One of	$\{x1 \ldots xn\}$	EquivalentProperty	$r1 \equiv r2$
AllValuesFrom	$\forall r.C$	DisjointWith	$C \sqsubseteq \neg D$
SomeValuesFrom	$\exists r.C$	SameAs	$\{x1\} \equiv \{x2\}$
HasValue	$\exists r.\{x1\}$	DifferentFrom	$\{x1\} \sqsubseteq \neg\{x2\}$
MinCardinality	$(\geq n\ r)$		
MaxCardinality	$(\leq n\ r)$		
InverseOf	r^{-}		

7.4 Composite Activity Modelling

7.4.1 Concept and Terminology

This section provides a set of definitions for concepts that are used to specify activity models in subsequent sections

7.4.1.1 Characterization of the Contextual Information of Smart Environments

To model smart environments, we identify and define the following sets and transformations between sets: environmental entities (O), sensors (S), sensor observations (SO), and associated context information (C).

Definition 1 The set of all sensors, S, lists all physical sensors installed in the environment. It is defined in (7.1).

$$\text{S}: \{\text{s1}, \text{s2}, \ldots, s_q\} \tag{7.1}$$

Definition 2 The set of all possible sensor observations, SO, lists all sensor observations that are made in the environment. Each physical sensor can generate one or more sensor observations over time. It is defined in (7.2).

$$\text{SO}: \{\text{so1}, \text{so2}, \ldots, so_z\} \tag{7.2}$$

Definition 3 The set of all objects, O, lists all objects that the user can interact with in the smart environment. It is defined in (7.3).

$$\text{O}: \{\text{o1}, \text{o2}, \ldots, o_m\} \tag{7.3}$$

Definition 4 The set of all context elements, C, lists all context elements that are monitored during activity recognition. For example, it can include temporal or spatial context. It is defined in (7.4).

$$\text{C}: \{\text{c1}, \text{c2}, \ldots, c_p\} \tag{7.4}$$

Definition 5 The function, f, maps a sensor observation to the corresponding object that the user just interacted with. By iteratively applying the function, f, to the set of sensor observations, the list of objects that the user has interacted with in a given time interval can be derived and used to describe a user activity. It is defined in (7.5).

$$\text{f}: \text{soi} \rightarrow \text{oj}, \text{soi} \in \text{SO}, \text{oj} \in \text{O} \tag{7.5}$$

7.4.1.2 Characterization of Activities of Daily Living

To help understand and characterize the human activities, we introduce various terminologies, namely, action, activity description, simple activity, and static and dynamic composite activities. These terms are used to derive composite activity models in the next section.

Definition 6 Primitive action (a): A single indivisible activity performed by the user. A primitive action is specified as a 2-tuple consisting of a collections of sensor observations and context information as provided in (7.6).

$$a: \langle \text{SO}_\text{a}, \text{CO}_\text{a} \rangle, \text{SO}_\text{a} \subseteq \text{SO}, \text{C}_\text{a} \sqsubseteq \text{C} \tag{7.6}$$

Definition 7 Activity description (AD): A collection of primitive actions, ai, over a specific time interval. An activity description may fully or partially describe an activity and is specified using a set as shown in (7.7).

$$\text{AD}: \{a_1, a_2, \ldots, a_m\} \tag{7.7}$$

Definition 8 ADL: This is the set that lists all activities of daily living (ADL) concepts, A_n, for defining simple activities in the activity model and is specified in (7.8).

$$\text{ADL}: \{A_1, A_2, \ldots, A_n\} \tag{7.8}$$

Definition 9 IADL: This set provides a list of all leaf ADL concepts, i.e., ADL concepts with no child concepts. It is defined in (7.9).

$$\text{IADL}: \{IA_1, IA_2, \ldots, IA_k\}, \text{k} < n, IADL \sqsubseteq ADL \tag{7.9}$$

Definition 10 Simple Activity (lAi): An ordered sequence of primitive actions. It is specified in (7.10).

$$\mathrm{IA}_i :< \mathrm{ADL}, \mathrm{L} > \tag{7.10}$$

where L is a text string to act as the label for the ongoing activity and ADL is an activity description for activity L.

Definition 11 Composite activity: A collection of two or more simple activities occurring within a given time interval.

Definition 12 Dynamic composite activities (dCA) set: lists a collection of all sequential, or interleaved and concurrent activities. It is specified in 7.11.

$$\mathrm{dCA}: \{\mathrm{dcA}_1, \mathrm{dcA}_2, \ldots, \mathrm{dcA}_d\} \tag{7.11}$$

Definition 13 Single dynamic composite activity (dcA_i): A composite activity that has properties whose values vary in time, implying the notion of change. It is defined in (7.12).

$$\mathrm{dcA}_i\!: \; < \phi, \tau, L >, dcA_i \in dCA_a \tag{7.12}$$

where L is a text string to act as a label for the pattern and Φ is a collection of leaf ADLs or a collection of dynamic composite activities such that $\phi \sqsubseteq \mathrm{IADL} \cup \mathrm{dCA}$. In addition, τ a subset of C, is the union of temporal contexts for all activities in dcA_i.

To illustrate a dynamic composite activity, consider the activity labelled 'make dinner and watch television'. We can specify Φ as $\Phi = \{$make dinner, watch television$\}$. In addition, τ can be specified by $\tau = \{$time-interval-of-make-dinner, time-interval-of-watch-television$\}$.

Definition 14 Static composite activity (sCA) set: defines a set of all sequential, or interleaved and concurrent activities as shown in (7.13).

$$\mathrm{sCA}: \left\{\mathrm{scA}_1, \mathrm{scA}_2, \ldots, \mathrm{scA}_g\right\} \tag{7.13}$$

Definition 15 A single static composite activity (scAi): This is a composite activity whose properties take values that do not change in time. It is specified as a 3-tuple in (7.14).

$$\mathrm{scA}_i\!: \langle \phi, \theta, L \rangle, scA_i \in sCA \tag{7.14}$$

where, ϕ is a collection of leaf ADLs or a collection of static composite activities such that $\phi \sqsubseteq \mathrm{IADL} \cup \mathrm{sCA}$. θ is an aggregate of task contexts associated with contained activities and it is a subset of C.

To illustrate a static composite activity, given the activity 'make dinner and watch television', we have $\phi = \{make\ dinner, watch\ television\}$. Also, θ is specified by task context given by descriptions, i.e., '*make dinner begins*'; '*as make dinner continues watch television begins*'; '*make dinner continues and ends*' and the relationship is parallelism.

7.4.2 Ontological Composite Activity Modelling

Based on the definitions in the previous sub-section and the revised conceptual activity model, we have developed ontological concepts for specifying simple and composite activity models which are discussed. Table 7.3 lists the properties of these concepts and Table 7.4 displays a fragment of the DL formulas for selected concepts.

7.4.2.1 Ontological Modelling of Activities, Temporal and Environmental Context

- *Activity*: This concept is the overall concept for all types of activities.
- *MonitoredEntity*: This is a general concept to represent the set of all entities in the *environment* occupied by an actor. Each *MonitoredEntity* relates to sub-concepts of Activity using the *hasMonitoredEntity* property.
- *Location*: This is a context concept that is used to indicate the location of interest to the actor. For instance, in a SH environment, it can be used to represent locations like the kitchen, bedroom, bathroom, living room, lounge, etc. Each Location concept can relate to the *MonitoredEntity* and Activity concepts using the *hasLocation* property.
- *Sensor*: This concept is used to denote the class of all sensors that are deployed in a smart environment. To link sensors to environment entities and objects, instances of Sensor are associated with instances of *MonitoredEntity* using the *hasSensor* property. Instances of the Sensor concept are used to encode the runtime state of the SH environment when the user is performing activities. Such runtime information can be used to derive contextual information.
- *TimeInterval*: This concept defines a time interval and indicates the moment of time that a time slice refers to.
- *TimeSlice*: This encodes the temporal extent of activities as a collection of time slices. The temporal extent is specified by associating Activity concept with *TimeSlice* using *timeSliceOf* property. In addition, instances of *TimeSlice* relate to instances of *TimeInterval* through the *hasTimeInterval* property.

Table 7.3 Properties for the concepts in the activity models

Name	Domain	Range	Other properties	Description
`hasMonitored Entity`	`ADLActivity`	`MonitoredEntity`		Indicates the entities used by an ADLActivity
`hasLocation`	`Activity, MonitoredEntity`	`Location`		Associates activity and MonitoredEntity with their spatial context.
`hasSensor`	`MonitoredEntity`	`Sensor`		Attaches environment entities to be monitored to the respective sensors.
`timeSliceOf`	`TimeSlice, BasicActivityTS, CompositeActivityTS`	`StaticComposite Activity, ADLActivity, DynamicComposite Activity`	`Functional`	Indicates temporal extent of activity concepts
`hasTimeInterval`	`TimeSlice`	`TimeInterval`	`Functional`	Associates TimeSlice to TimeInterval
`hasOngoing Activity`	`CompositeActivityTS`	`ADLActivityTS, CompositeActivityTS`	`Irreflexive`	A dynamic (fluent) property that captures the notion of change.
`hasActivity`	`StaticCompositeActivity`	`ADLActivity, StaticCompositeActivity`	`Irreflexive`	Provides components of a composite activity
`startedBy`	`StaticCompositeActivity`	`ADLActivity, StaticCompositeActivity`	`Irreflexive`	Indicates starting activity of a composite activity
`endedBy`	`StaticCompositeActivity`	`ADLActivity, StaticCompositeActivity`	`Irreflexive`	Indicates activity that marks the end of composite activity.
`Entails Composite Activity`	`StaticCompositeActivity`	`DynamicCompositeActivity`		Used to indicate if a given DynamicCompositeActivity has a corresponding StaticCompositeActivity in the model
`relationshipType`	`StaticCompositeActivity`	`String`	`Functional`	Indicates whether a SEQUENCE or PARALLEL relation exists

Table 7.4 DL formulas for a select set of concepts used in activity models

• $MonitoredEntity \sqsubseteq \exists hasSensor.Sensor \sqcap \exists hasLocation.Location$
• $Activity \sqsubseteq \exists hasLocation.Location \sqcap (ADLActivity \sqcup CompositeActivity)$
• $ADLActivity \sqsubseteq Activity \sqcap \exists hasMonitoredEntity.MonitoredEntity$
• $TimeSlice \sqsubseteq \exists timeSliceOf.Activity \sqcap = 1 timeSliceOf \sqcap \exists hasTimeInterval.Interval \sqcap = 1 hasTimeInterval$
• $BasicActivityTS \sqsubseteq TimeSlice \sqcap \exists timeSliceOf.ADLActivity \sqcap = 1 timeSliceOf$
• $CompositeActivityTS \sqsubseteq TimeSlice \sqcap \exists timeSliceOf.DynamicCompositeActivity \sqcap = 1 timeSliceOf \sqcap \exists hasOngoingActivity.TimeSlice \sqcap \geq 2 hasOngoingActivity$
• $StaticCompositeActivity \equiv \exists hasActivity.(ADLActivity \sqcup StaticCompositeActivity) \sqcap \geq 2 hasActivity \sqcap \exists startedBy.(ADLActivity \sqcup StaticCompositeActivity) \sqcap \exists endedBy.(ADLActivity \sqcup StaticCompositeActivity) \sqcap \exists entailsCompositeActivity.DynamicCompositeActivity \sqcap \exists relationshipType.string$

7.4.2.2 Concepts for Simple Activities

- *ADLActivity*: This concept is the parent concept to all simple activity concepts. All simple activities are defined as subclasses of the *ADLActivity* concept. In general, given that the hypothetical ADL, *SimpleADL*, is a subclass of *ADLActivity*, it can be declared in DL as:

$$SimpleADL \sqsubseteq ADLActivity \sqcap \exists property1.Range \sqcap \exists property2.Range2 \ldots \sqcap \exists propertyN.RangeN \tag{7.15}$$

- In the above example, *property1...propertyN* are sub-properties of *hasMonitoredEntity*, and *Range1...RangeN* are sub-concepts of *MonitoredEntity* associated with *SimpleADL*. The sensor observations and context specified in Definitions 2 and 3 are realised by property restrictions that are defined on *ADLActivity*. For example, a typical observation, e.g. '*using the kettle*' can be represented by a *hasKettle* property restriction that is defined on the relevant subclasses of *ADLActivity*. In this example, *hasKettle* is a sub-property of *hasMonitoredEntity*; whereas the concept for kettle is a sub-concept of *MonitoredEntity*. Further details on simple activity concepts will be provided in Sect. 7.5.1. since the concepts are dependent on the application domain.
- *BasicActivityTS*: This is a sub-class of *TimeSlice* and is used to add a temporal dimension to instances of *ADLActivity*.

7.4.2.3 Concepts for Composite Activities

This section defines various entities and terms, and their relationships for representing composite activities, which is depicted in Fig. 7.2.

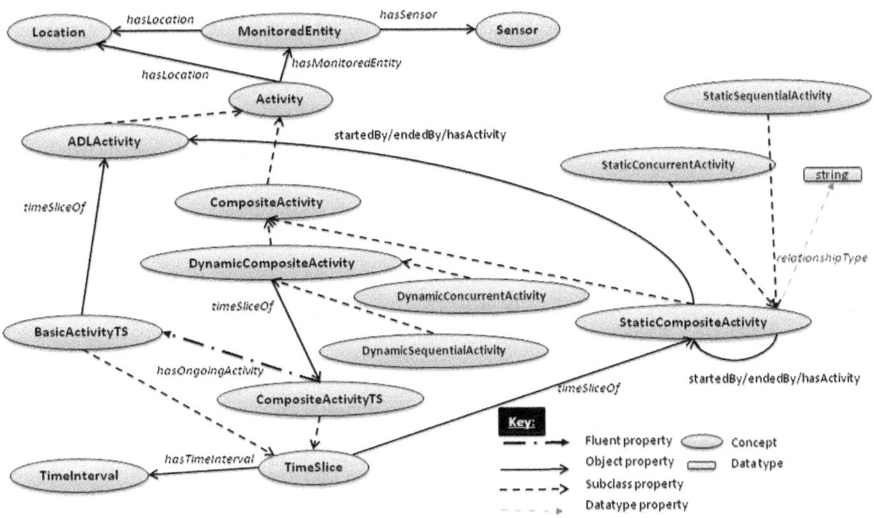

Fig. 7.2 A fragment of the activity models showing concepts and their inter-relationships

- *CompositeActivity*: This concept represents a dynamic or static composite activity. It is used to denote sequential and interleaved or concurrent activities.
- *DynamicCompositeActivity*: This is a sub-concept of *CompositeActivity* and represents a dynamic composite activity corresponding to Definition 13. Property restrictions to encode change are defined on this concept. For instance, the notion of change is represented by implications derived from the fluent property *hasOngoingActivity*. The instances of this concept are intended to be derived at runtime.
- *CompositeActivityTS*: This is a subclass of *TimeSlice* that relates to *DynamicCompositeActivity*. It explicitly captures the notion of change by defining a restriction on the fluent property *hasOngoingActivity*. Essentially, objects of *CompositeActivityTS* associate with objects of *BasicActivityTS* or another *CompositeActivityTS* through the fluent property *hasOngoingActivity* to denote change. Each composite activity can be derived from the activities whose *TimeSlice* objects have been associated with the *hasOngoingActivity* over a given time interval. It associates with *DynamicCompositeActivity* using the *timeSliceOf* property.
- *DynamicConcurrentActivity*: Sub-concept of *DynamicCompositeActivity* whose instances are dynamic concurrent or interleaved activities.
- *DynamicSequentialActivity*: Sub-concept of *DynamicCompositeActivity* whose instances are dynamic sequential activities.
- *StaticCompositeActivity*: This is a sub-concept of *CompositeActivity* and defines a static composite activity as per Definition 15. It simply captures the activities (whether simple or composite) that constitute a composite activity. It typically captures inter-activity relations using the *hasActivity* property to specify the activities that constitute the composite activity. It associates with *DynamicCompositeActivity* through the *entailsCompositeActivity* property. This concept can specify a

time slice but does not encode the notion of change, hence, it relates to *TimeSlice* concept using *timeSliceOf* property.
- *StaticConcurrentActivity*: Sub-concept of *StaticCompositeActivity* whose instances are static concurrent or interleaved activities.
- *StaticSequentialActivity*: Sub-concept of *StaticCompositeActivity* whose instances are static sequential activities.

7.4.3 *Interval Temporal Logic in Composite Activity Modelling*

In addition to ontological modelling of relationships between activities and entities described above, we use Allen interval relations to model temporal relationships between simple activities of composite activities. The models show how to relate temporal intervals of composite activities to the temporal intervals of their composing activities. The resulting models can be used to infer composite activities from temporal intervals of simple activities or other composite activities.

7.4.3.1 Models of Sequential Composite Activities

Sequential activities are modelled by associating their respective intervals using the Allen relations before/after and meets/met-by. The relation before/after signifies that there is a gap between the two intervals, while meets/met-by indicates that the two intervals follow each other with no gap between them. These two relations (marked by solid arrows) and their implications (marked by dotted arrows) are represented in Fig. 7.3a.

7.4.3.2 Models of Interleaved and Concurrent Composite Activities

The models of interleaved and concurrent activities encode the notion that activities can occur simultaneously only if their time intervals overlap fully or partially. The models of interleaved and concurrent activities are created using nine temporal relations, i.e., overlaps/overlapped-by, during/contains, starts/started-by, finishes/finished-by and equals as described below. The resulting temporal models are shown in Fig. 7.3a–f:

a. *Before-/after* – show activity two occurring after activity one has finished.
b. *Overlaps/overlapped-by*- this shows that two activities have components of their intervals that are shared, but with one interval starting or ending before the other interval.

c. *Contains/during*- this model a composite activity made up of simple activities, e.g. 'prepare meal' that contains 'prepare soup' and 'prepare vegetable'. The longer interval encloses the shorter one;
d. *Starts/started-by*-shows the simple/composite activity that starts another simple/composite activity.
e. *Finishes/finished-by*- shows the simple/composite activity that finishes another simple/composite activity.
f. *Equals*- Theoretically, this scenario only applies to concurrency by parallelism. The two intervals start and end at the same time.

7.4.3.3 Entailment Rules for Activity Modelling

The previous section describes interval-based temporal interrelationships between simple activities of a composite activity using ontological concepts. Nevertheless, the mechanism for interpreting the temporal relationships is still missing, which is required in order to infer composite activities. To this end, we have defined a set of entailment rules as an essential part of the composite model, based on Semantic Web Rule Language (SWRL) [29], which can infer complex dependencies among

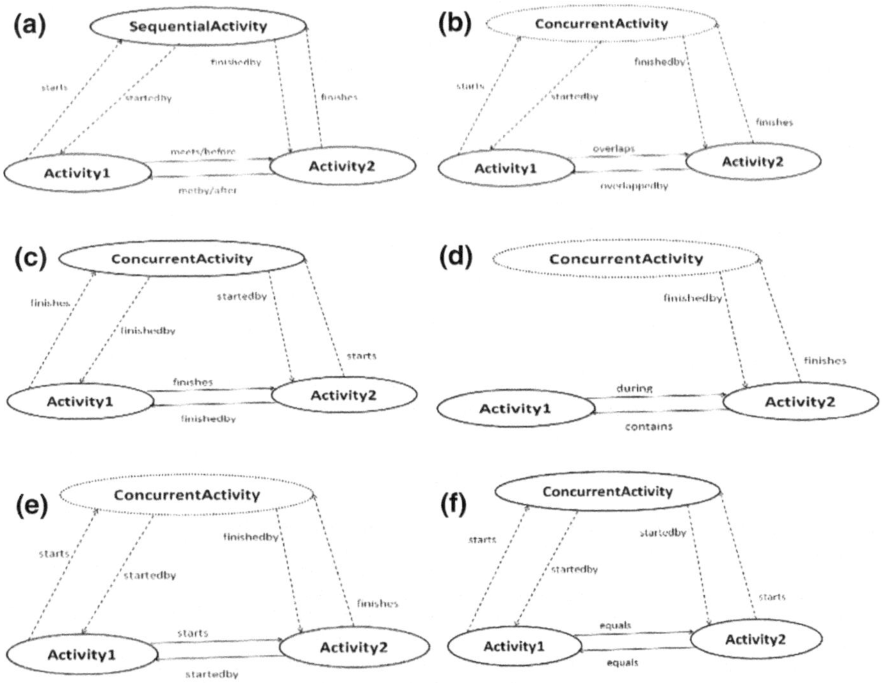

Fig. 7.3 Temporal relationship models of composite activities

activities and therefore the ongoing composite activities. These rules can be used to derive the ongoing composite activities by identifying the existing relationships between temporal intervals of ongoing activities. Three categories of entailment rules have been designed, namely, rules to derive interval relations and assert dynamic composite activities; rules to assert instances of fluent property; and rules to derive and assert static composite activities. Due to limitation of space, in the following we use three entailment rules based on the overlaps/overlapped-by relationship to illustrate the development of rules and their use for activity inference.

7.4.3.4 Derive Interval Relations and Assert Dynamic Composite Activity Intervals

Given two existing intervals for a pair of primary activities that have a qualitative temporal relationship, the rules in Table 7.5a, b are used to assert interval end-points of a dynamic composite activity. The rule in Table 7.5a derives the interval relation *intervalOverlaps* by using the interval end-points of two primary activities. Table 7.5b provides the rule for obtaining the inferred values of *intervalOverlaps* property. The left-hand side (LHS) of the rule in Table 7.5b obtains three *TimeSlice* objects and determines the beginning and ending points for each primary activity's interval. Finally, it derives the existing interval relationship for the primary activities. The right-hand side (RHS) of the rule uses the facts established on LHS to assert the beginning and ending points of the time interval for the dynamic composite activity.

7.4.3.5 Assert Instances of Fluent Property

This is based on Table 7.5b, and the rule allows the *TimeSlice* objects linked to the ongoing primary activities to be related with the *TimeSlice* object of the *DynamicCompositeActivity* through the fluent property, *hasOngoingActivity*. The LHS obtains three *TimeSlice* objects, i.e., the two for primary activities and one for the dynamic composite activity, checks for the temporal dependency between the primary activities, and asserts instances of the fluent property. If two *TimeSlice* objects share a temporal relation, then they are associated with the *TimeSlice* object of the dynamic composite activity using the fluent property.

7.4.3.6 Derive and Assert Static Composite Activities

This is based on Table 7.5c above and the rule's LHS checks that there exists an instance of *DynamicCompositeActivity*, an instance of *StaticCompositeActivity*, as well as instances of the fluent property that are defined on the former's instance. The RHS then uses the facts established by the LHS to assert instances of the *entailsCompositeactivity* that is defined as a property of the concept *StaticCompositeActivity* (see Table 7.3). Essentially, this rule is used to infer and validate the ongoing composite activity that is subsequently reported to the user. Validation fails if instances of *DynamicCompositeActivity* do not have corresponding instances of *StaticCompositeActivity*. Whenever validation fails the resulting composite activity described by the instance of *DynamicCompositeActivity* should be suggested for addition into the static model of activities.

Similar rules are defined for other qualitative temporal relations (i.e., equals, during/contains, starts/started-by, finishes/finished-by, before/after, and meets/met-by) but due to space limitations we do not provide them here.

Table 7.5 Entailment rules for inferring composite activities

```
(a) ProperInterval(?x), ProperInterval(?y), before(?a,?c),
    before(?b,?d), before(?c,?b), hasBeginning(?x,?a),
    hasBeginning(?y,?c), hasEnd(?x,?b), hasEnd(?y,?d) ->
    intervalOverlaps(?x,?y)
(b) ADLActivity(?ax), ADLActivity(?ay),
    ComplexActivityTS(cplxConcurrentTS),
    DynamicConcurrentActivity(dynConcurrentActivity),
    TimeSlice(?tsx), TimeSlice(?tsy),
    Interval(dynConcurrentInterval), ProperInterval(?x),
    ProperInterval(?y), hasTimeInterval(?tsx,?x),
    hasTimeInterval(?tsy,?y),
    hasTimeInterval(cplxConcurrentTS,?w), timeSliceOf(?tsx,?ax),
    timeSliceOf(?tsy,?ay), timeSliceOf(cplxConcurrentTS,
    dynConcurrentActivity), hasBeginning(?x,?a), hasEnd(?y,?d),
    intervalOverlaps(?x,?y) ->
    hasOngoingActivity(cplxConcurrentTS,?tsx),
    hasOngoingActivity(cplxConcurrentTS,?tsy),
    hasBeginning(?w,?a), hasEnd(?w,?d), intervalFinishes(?y,?w),
    intervalStarts(?x,?w)
(c) ComplexActivityTS(?tsw), ConcurrentActivity(?sa),
    DynamicConcurrentActivity(?aw), TimeSlice(?tsx),
    TimeSlice(?tsy), endedBy(?sa,?ay), hasActivity(?sa,?ax),
    hasActivity(?sa,?ay), hasOngoingActivity(?tsw,?tsx),
    hasOngoingActivity(?tsw,?tsy), hasTimeInterval(?tsw,?w),
    hasTimeInterval(?tsx,?x), hasTimeInterval(?tsy,?y),
    startedBy(?sa,?ax), timeSliceOf(?tsw,?aw),
    timeSliceOf(?tsx,?ax), timeSliceOf(?tsy,?ay),
    intervalFinishes(?y,?w), intervalOverlaps(?x,?y),
    intervalStarts(?x,?w), relationshipType(?sa, "PARALLEL") ->
    entailsCompositeActivity(?aw,?sa).
```

7.5 Simple and Composite Activity Recognition Methods

7.5.1 Ontological and Temporal ADL Models

To support SH-based activity recognition, we have aggregated the models created above including concepts and properties for simple and composite activities, and the entailment rules to form the ADL Ontology. The ADL ontology is constructed based on common ADL activities related to food preparation, recreation, and *hygiene*. Table 7.6 shows a list of simple ADL activity models, namely ADL concepts, their properties, and the range concepts representing the objects used by the users to perform respective ADLs. All the Range concepts are sub-concepts of *MonitoredEntity*. The super concepts are listed in the order from the immediate super-concept to the most general super-concept. These simple activities can form composite activities if a user performs them in sequence or concurrently, as discussed in Sect. 7.6

7.5.2 Composite Activity Recognition Architecture

Whenever a user performs activities along a timeline, the sensor data stream can be analysed to recognize the ongoing activities based on the extracted contextual information. A sensor data stream can be partitioned to obtain segments of sensor data and associated contextual information. To partition a sensor data stream, the time window-based segmentation method described in Chap. 5, has been used to support dynamic segmentation. The segmentation approach uses information from the activity ontologies, e.g. activity duration, and feedback from activity inference. It generates segments from streaming sensor data that can be further analysed to infer the ongoing activities.

We propose the architecture shown in Fig. 7.4 to support both simple and composite activity recognition. The architecture supports three main tasks, namely, activity modelling, activity recognition, and sensor data stream segmentation. In activity modelling, two types of models (i.e., static and dynamic models as previously described) are created as described in Sect. 0. To create the two models, activity modelling involves two processes, i.e., static activity modelling and dynamic activity modelling. Static activity modelling involves creating static models of simple and composite activities. Dynamic activity modelling is used to create the dynamic model of composite activities. Activity recognition is modelled as three interdependent tasks, namely, action recognition, simple activity recognition, and composite activity recognition. The action recognition task is tightly coupled with and subsumed in the simple activity recognition task. By separating activity recognition into interdependent tasks, it is possible to use different techniques for each task. In this work, instance retrieval, subsumption reasoning and equivalence reasoning are used for action and simple activity recognition. For composite activity recognition, rule-based inference techniques are exploited. Finally, segmentation is used to aid

Table 7.6 Summary of ADL concepts in the ADL ontology

Concept	Super concept	Property	Range concepts
Have bath	BathRoomADL, BasicADLActivity, ADLActivity, Activity	hasHygieneItem	BathBrush, BathSoap, BodyWash, Towel
		hasHygieneAppliance	BathTap
		hasLocation	BathRoom
Brush teeth	BathRoomADL, BasicADLActivity, ADLActivity, Activity	hasHygieneAppliance	Toothbrush, WashingSinkTap
		hasHygieneItem	Toothpaste, MouthWash
		hasLocation	BathRoom
Wash hands	BathRoomADL, BasicADLActivity, ADLActivity, Activity	hasHygieneAppliance	WashingSinkTap
		hasHygieneItem	Towel, HandWash
		hasLocation	BathRoom
Make tea	MakeHotDrink, MakeDrink, KitchenADL, FunctionalADLActivity, ADLActivity, Activity	hasFlavoring	Milk, Sugar
		hasContainer	Cup
		hasHotDrinkType	Tea
		hasLocation	Kitchen
Make chocolate	MakeHotDrink, MakeDrink, KitchenADL, FunctionalADLActivity, ADLActivity, Activity	hasFlavoring	Milk, Sugar
		hasContainer	Cup
		hasHotDrinkType	Chocolate
		hasLocation	Kitchen
Make coffee	MakeHotDrink, MakeDrink, KitchenADL, FunctionalADLActivity, ADLActivity, Activity	hasFlavoring	Milk, Sugar
		hasContainer	Cup
		hasHotDrinkType	Tea
		hasLocation	Kitchen
Make pasta	MakeHotMeal, MakeMeal, KitchenADL, FunctionalADLActivity,ADLActivity, Activity	hasFlavoring	Salt
		hasContainer	Plate
		hasCookingAppliance	Cooker
		hasHygieneAppliance	KitchenSinkTap
		hasMaterial	Pasta
		hasUtensil	Drainer, Pan
		hasLocation	Kitchen
Watch television	LoungeADL, RecreationalADLActivity, ADLActivity, Activity	hasEntertainmentAppliance	TV, TVRemote
		hasFurniture	Sofa
		hasLocation	Lounge

real-time processing by supporting online segmentation of streaming sensor data. The resulting segments are processed during activity recognition and mapped to corresponding simple or composite activities.

7.5.3 Composite Activity Recognition Algorithm

Given a segment of sensor data stream, we describe the algorithm that derives the corresponding simple or composite activities. The steps described here are summarized in the algorithm listing in Table 7.7.

In the first step, the enclosed observations are converted into primitive actions by checking property restrictions specified in the ADL Ontology through ontological reasoning. The second step groups primitive actions into one or more activity descriptions corresponding to the simple activities that are defined in the ADL Ontology. Each activity description is constructed such that it contains only sensor data and context information that can be mapped to a single specific simple activity or to a general class of simple activities. For instance, an activity description can match the definition of the general class '*MakeHotDrink*' (a super activity for '*MakeTea*, '*MakeCoffee*', and '*MakeChocolate*') or that of the simple class '*MakeTea*'. Therefore, an activity description is considered more general if it corresponds to a general class of activities and more specific otherwise. A more general activity description signifies that more sensor data and context information is still needed to identify the correct sub-activity.

To group primitive actions, the algorithm checks the ontology to determine the ADL concepts that are in the domain of the corresponding property restriction. For instance, given the knowledge that the user is in the kitchen, the algorithm can obtain

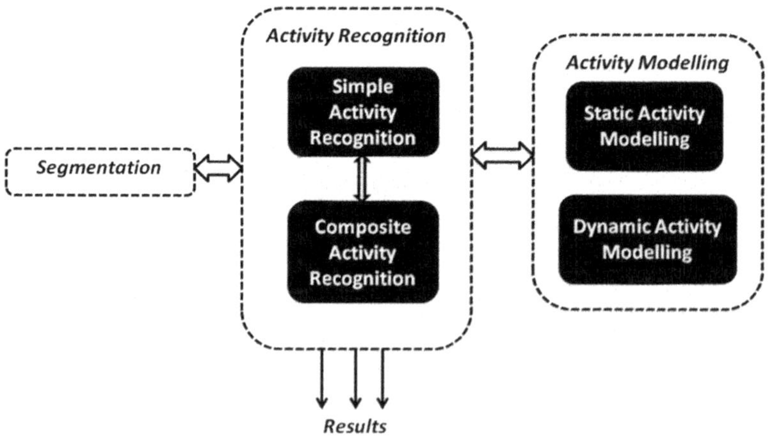

Fig. 7.4 A logical architecture for composite activity modelling and recognition

Table 7.7 Temporal reasoning algorithm for composite activity recognition

```
INPUT: sensor data stream (Ω), ADL ontology (ADL-O), inference rule base (RB)
OUTPUT: Composite activity (CA) or simple activity (SA)
RECOGNIZE-ACTIVITY (Ω, ADL-O, RB)
BEGIN:
  WHILE data stream is active DO
     Segment data stream (Ω) into a set of segments S={s_1, s_2, ..., s_n}
     /*for each segment map observations to activities*/
     FOR each s_i∈S DO
        /*Create activity descriptions to form set AD={AD_1, AD_2,...AD_k}*/
        Extract observations from segment s_i, O={o_1,o_2,...o_h}
        FOR each o_i∈O DO //for each observation
           Retrieve all activities described by o_i,  A(o_i)={A_1, A_2, ...A_n}
        ENDFOR
        Create a set of all activities A=A(o_1)⊔ ...A(o_i)⊔ ... A(o_h)
        FOR each x∈A DO/*for all activities*/
           Collect all observations constituting the definition of x as activity descriptions AD_x
           Add AD_x to AD
        ENDFOR
        FOR each AD_j∈AD DO /*classify activity*/
           Map AD_i to a simple activity /*use ontological inference*/
           IF a leaf activity is returned THEN
              Report it (SA)
           ELSE
              IF goal is still valid THEN
                 Wait for updated activity description (go to start of current loop)
              ELSE Communicate status report and terminate
              ENDIF
           ENDIF
           Update classification status into recognition status RS= {st_1, st_2, ... st_m}
        ENDFOR
        /*to help aggregate results for composite activities*/
        Define the set of time intervals TI
        FOR each st_i ∈ RS DO Obtain temporal interval I_i and add it to TI ENDFOR
        IF only one interval is present THEN
           Report it (SA) /*SA-activity in interval I*/
        ELSE
           Infer interval relations using RB
           Derive ongoing composite activity relationships
           Check corresponding instances of static composite activities
           IF instance in static model THEN Report it (CA)
           ELSE Recommend the activity ontology be updated to accommodate it
           ENDIF
        ENDIF
        IF results are conclusive THEN Convey results (SA or CA)
        ELSE Update segment and window ENDIF
     ENDFOR
   ENDFOR
 ENDWHILE
 END
```

the triples < *MakeTea, hasLocation, kitchen* > , < *MakeCoffee, hasLocation, kitchen* > , and < *MakeChocotate, hasLocation, kitchen* > , and so on, for all the activities that can take place in the kitchen. As a result, it will form a number of partial activity descriptions with one description per possible activity and each activity will be associated with the primitive action representing 'in the kitchen'. Since each activity description contains only sensor data and context information corresponding to an activity or class of activities, several activity descriptions can be generated for a given segment of the sensor data stream.

As more sensor data is obtained new activity descriptions will be created or the existing ones will be modified and enriched. For example, if the next sensor observation relates to the user using the tea bag, only the partial activity description corresponding to *MakeTea* will be updated to accommodate the new primitive action. During this process, a few partial activity descriptions will become complete when all the relevant primitive actions have been executed. However, all the existing partial or complete activity descriptions will remain valid until sensor observations within a given segment are discarded.

In the third step, simple activity recognition is performed to map activity descriptions into activity labels. To perform simple activity recognition, we adopt and modify the ontological reasoning approach described in Chaps. 3 and 4. The original approach progressively aggregates sensor data within a data stream segment and performs subsumption and equivalence reasoning to infer the entailed activity. In the modified approach, the algorithm processes each activity description obtained as previously described against the ADL Ontology. Basically, it compares each activity description with activity models in the ADL Ontology using semantic reasoning and the activity label for the model that is closest to the activity description is reported as the ongoing simple activity. The activity model returned by instance retrieval is considered the closest model. In the absence of a model returned by instance retrieval, then the model returned by equivalence reasoning is taken as the closest. Otherwise, the model returned by subsumption is the closest. Given the possibility of multiple activity descriptions per data stream segment, parallel simple activity recognition processes can be initiated with each process dedicated to a single activity description.

The fourth and last step performs composite activity recognition by using the inference rules to aggregate the results of simple activity recognition. The strategy is to progressively aggregate the results of simple activity recognition in order to recognize composite activities. If only one simple activity has been identified for a sensor data segment, this can be reported to the user. Alternatively, if more than one simple activity is identified from corresponding activity descriptions, the results are processed to determine if ongoing simple activities share qualitative temporal relationships. The simple activities that share qualitative temporal relationships are inferred as components of a composite activity. However, before the composite activity is reported to the relevant applications, the ontology is checked for a corresponding instance in the static activity model. If a corresponding instance exists, it is reported to the user; otherwise, it is considered a novel composite activity and recommended for inclusion in the ontology. To perform this analysis, the approach uses temporal inference rules. The rules can infer qualitative temporal relationships and derive corresponding composite activities from the activity models using the temporal inference mechanism.

7.6 An Example Case Study

7.6.1 System Prototype

As a case study for testing and evaluation, we developed the proposed approach into a system prototype for simple and composite activity recognition. The prototype consists of the ADL Ontology and a multi-agent system. The multi-agent system consist is built using Java Agents Development Framework (JADE) [30]. Overall, four types of agents were implemented as follows: (a) an agent to receive sensor input and to segment the sensor data stream; (b) an agent to manage action inference, generate activity descriptions, and to oversee summarization of activity information; (c) an agent to manage inference rule execution to aggregate the results from simple activity recognition to infer composite activities and, finally; (d) multiple agents to infer simple activities. The benefit of choosing agent as an implementation artefact is because agents provide the different components with autonomy needed to perform their respective tasks. In addition, each component can continuously review and react to changes in its goals. Also, there is massive parallelism involved in executing the various tasks involved in the framework and these tasks are implemented as agent behaviours. The ADL Ontology was built using OWL 2 [6] constructs in Protégé [31] ontology editor. The prototype implements Java-based code interacting with the Pellet [32] OWL reasoner to provide ontological reasoning support. The inference rules were implemented as Semantic Web Rule Language (SWRL) [29] rules. To facilitate the execution of the inference rules we translated the activity ontology and the SWRL rules to Java Expert System Shell (JESS) [33] fact and rule bases, respectively. We used the OWL2Jess and SWRL2Jess translators based on [34]. In the prototype, the JESS fact and rules bases are accessed and processed by a JESS rule engine. The engine is accessed by behaviours in the JADE agent responsible for aggregating the results of simple activity recognition. A segment of the activity ontology and a runtime snapshot of the agent system are shown in Fig. 7.5.

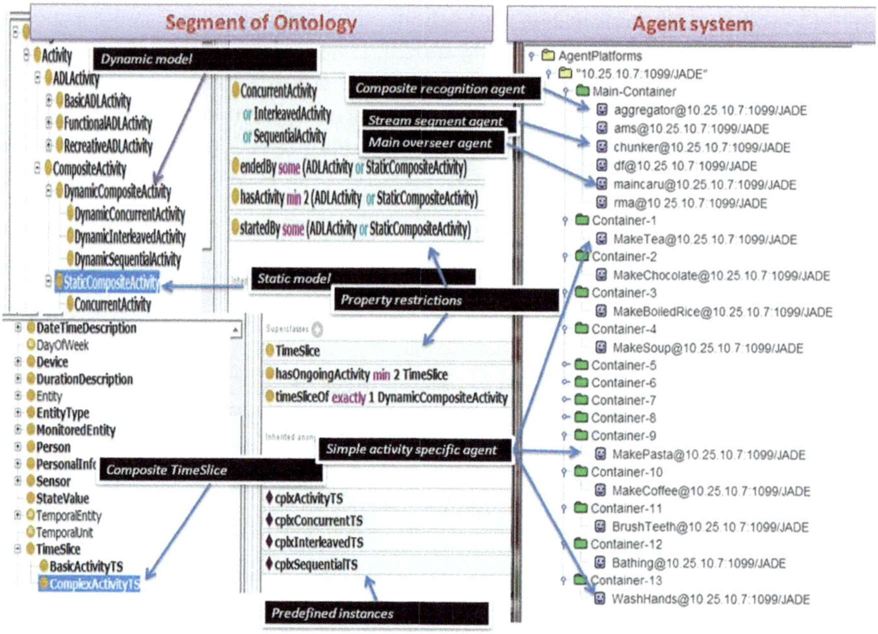

Fig. 7.5 A snapshot of the interactions between activity ontologies and runtime agent system

7.6.2 Experiment Design

To evaluate and demonstrate the feasibility of the proposed approach, we used the synthetic data generator developed and described in Chap. 5 (Sect. 5.6.1) to generate synthetic ADL data. To generate the synthetic ADL data, seven typical simple ADLs related to meals (e.g. *MakeTea*, *MakeCoffee*, *MakeChocolate*, and *MakePasta*), hygiene (e.g. *HaveBath*, *WashHands*) and recreation (e.g. *WatchTelevision*) were used. The synthetic ADL data possesses the necessary temporal information and allows us to evaluate the feasibility of the developed approach. To generate synthetic ADL data, we specified 'seed' ADL patterns for both simple activities (e.g. *MakeTea*, *HaveBath*) and composite activities (e.g. *MakePastaAndMakeTea*). The synthetic ADL data generator then derives different permutations of these patterns. To select the permutation to use, it uses a random number generator to guarantee fairness in pattern selection. The transition time (in seconds) between ADLs is specified for each ADL pattern. For example, the pattern WashHands-0, MakePastaAndMakeTea-600, implies that *WashHands* is the first ADL in the pattern, while the ADL *MakePastaAndMakeTea* will occur 600 s after *WashHands* is completed. As described in Chap. 5 (Sect. 7.6.1), one or more patterns of sensor activations is provided for each simple ADL with each sensor in the pattern activated after a specific amount of time. It is therefore possible to derive the approximate activity duration from the temporal information associated with the sensor activations. We generated eight weeks of

Table 7.8 Summary of simple activities in synthetic data set

Simple activity	#In parallel	#In sequential	#Standalone	Sub-total
MakeTea	8	18	0	26
MakeCoffee	1	0	2	3
MakeChocolate	1	0	6	7
MakePasta	18	7	0	25
HaveBath	8	7	0	15
WashHands	0	10	0	10
WatchTelevision	10	8	0	18
Total				104

Table 7.9 Summary of composite activities in synthetic data set

Concurrent and interleaved	# Of occurrences	Sequential	# Of occurrences
MakePasta and MakeTea (a)	3	MakePasta then HaveBath (g)	6
MakePasta and WatchTelevision (b)	5	MakeTea then WashHands (h)	4
MakePasta and HaveBath (c)	8	WashHands then MakeTea (i)	6
WatchTelevision and MakeTea (d)	5	MakeTea then WatchTelevision (j)	3
MakePasta and MakeChocolate (e)	1	WatchTelevision then MakeTea (k)	5
MakePasta and MakeCoffee (f)	1	HaveBath then MakePasta (l)	1
Total	23 (46 simple activities)	Total	25 (50 simple activities)

synthetic ADL data consisting of 56 episodes of simple or composite activities. A total of 104 activities were generated consisting of 23 interleaved and concurrent activities (46 simple activities), 25 sequential activities (50 simple activities), and eight standalone simple activities to provide the ground truth. Table 7.8 presents an analysis of all simple activities while Table 7.9 provides a summary of composite activities.

7.6.3 Results and Discussions

To test the approach and associated algorithms for activity recognition, we use the simulation tool presented in Chap. 5 to play back the synthetic ADL data described

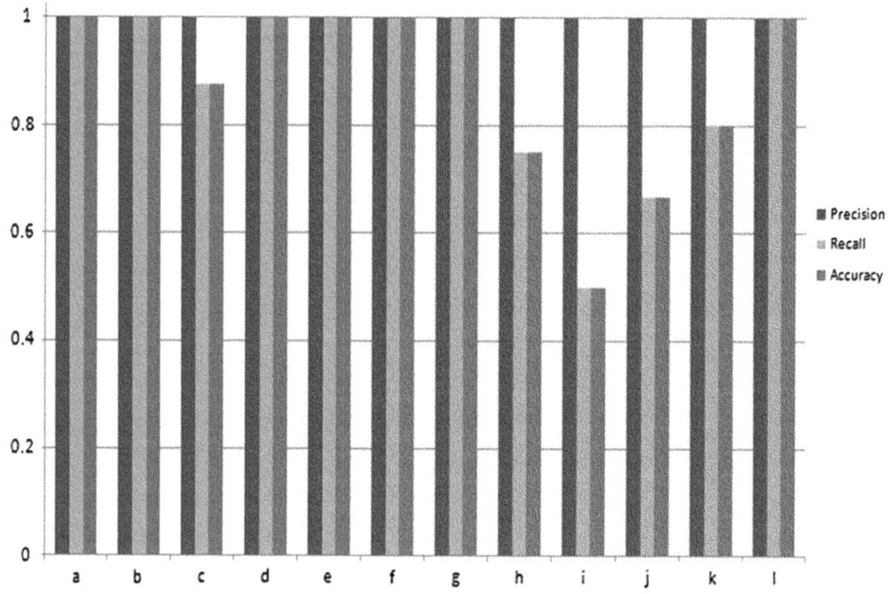

Fig. 7.6 Summary of results for composite activities

above. The sensor data is then fed to the activity recognition system as if the sensor activations are occurring in real-time. As the data is played back, the recognition system will attempt to identify the ongoing simple or composite activities. A total of 104 activities were played back in real-time and processed by the system prototype for activity recognition. The overall accuracy value obtained for simple activities is 100% since all 104 simple activities were successfully recognized. Figure 7.6 shows the precision, recall, and accuracy values for composite activities. An overall accuracy value of 88.26% was obtained.

We observed with interest that the accuracy for simple activities was 100%. We attribute this to the decision we made to derive all possible activity descriptions for each data stream segment. By deriving activity descriptions as presented by the algorithm in Table 7.7, the approach guarantees that only sensor observations that are relevant to a particular type of activity are included in its activity recognition. Essentially, primitive actions are used to derive activity descriptions before mapping the resulting descriptions to activity labels. As a result, each simple activity recognition unit can correctly classify its activity based on the sensor observations that it obtains.

The overall recognition accuracy for composite activities is 88.26% which is quite encouraging. It is important to note that recognition accuracy is lower when the composite activity consists of activities that are executed in sequence. This can be attributed to the transitions between activities and how well the system keeps track of the previously recognized activities. We believe that this can be increased by using feedback from composite activity recognition in segmentation. On the other hand, impressive accuracy results for concurrent and interleaved activities can be attributed to the absence of temporal transitions between the activities involved.

7.7 Summary

This chapter presented a hybrid approach and associated system architecture, models, methods and algorithms for composite activity modelling and recognition. The approach combines ontological and temporal knowledge representation formalisms to provide required modelling and representation capabilities for composite activity modelling. In this chapter, we have developed a generic conceptual activity model that encodes the characteristics of simple and composite activities and from which activity models can be specified. We have developed a unified activity recognition algorithm that processes streaming sensor data against composite activity models to support the identification of both simple and composite activities, e.g. interleaved and concurrent activities. We have described an integrated system architecture for composite activity recognition in smart homes and further developed a system prototype that was used to evaluate the approach. Using the prototype, we have conducted well-designed experiments which have observed average accuracy values of 100 and 88.26% for simple and composite activities, respectively. This research enriches the literature and advances the research frontiers of the knowledge-driven approach to activity recognition.

References

1. Modayil J, Bai T, Kautz H (2008) Improving the recognition of interleaved activities. In: Proceedings of the 10th international conference on Ubiquitous computing – UbiComp 2008
2. Patterson DJ, Fox D, Kautz H, Philipose M (2005) Fine-grained activity recognition by aggregating abstract object usage. In: Proceedings of international symposium on wearable computers, ISWC
3. van Kasteren T, Noulas A, Englebienne G, Kröse B (2008) Accurate activity recognition in a home setting. In: Proceedings of the 10th international conference on Ubiquitous computing - UbiComp 2008
4. Wu T, Lian C, Hsu JYY (2007) Joint recognition of multiple concurrent activities using factorial conditional random fields. In: Proceedings of 22nd conference artificial intelligence
5. Hao DH, Pan SJ, Zheng VW, Liu NN, Yang Q (2008) Real world activity recognition with multiple goals. In: Proceedings of the 10th international conference on Ubiquitous computing – UbiComp 2008
6. Hu DH, Yang Q (2008) CIGAR: Concurrent and interleaving goal and activity recognition. In: AAAI conference on artificial intelligence
7. Helaoui R, Niepert M, Stuckenschmidt H (2011) Recognizing interleaved and concurrent activities: A statistical-relational approach. In: 2011 IEEE international conference on pervasive computing and communications. PerCom 2011
8. Helaoui R, Niepert M, Stuckenschmidt H (2011) Recognizing interleaved and concurrent activities using qualitative and quantitative temporal relationships. In: Pervasive and mobile computing

9. Steinhauer H, Chua S (2010) Utilising temporal information in behaviour recognition.In: AAAI Spring Symposium
10. Okeyo G, Chen L, Wang H, Sterritt R (2012) A hybrid ontological and temporal approach for composite activity modelling.In: Proceedings of 11th IEEE international conference on trust, security and privacy in computing and communications trust. - 11th IEEE international conference ubiquitous computing and communication. IUCC-2012, pp. 1763–1770
11. Chen L, Nugent CD, Wang, H (2012) A knowledge-driven approach to activity recognition in smart homes. IEEE Trans Knowl Data Eng
12. Riboni D, Bettini C (2011) OWL 2 modeling and reasoning with complex human activities. Pervasive Mob, Comput
13. Chen L, Nugent C (2009) Ontology-based activity recognition in intelligent pervasive environments. Int J Web Inf Syst
14. Allen JF (2013) Maintaining knowledge about temporal intervals. In: Readings in qualitative reasoning about physical systems
15. Gu T, Wang L, Wu Z, Tao X, Lu J (2011) A pattern mining approach to sensor-based human activity recognition. IEEE Trans Knowl Data Eng
16. Saguna S, Zaslavsky A, Chakraborty D (2011) Recognizing concurrent and interleaved activities in social interactions. In: Proceedings IEEE 9th international conference on dependable, autonomic and secure computing, DASC 2011
17. Padovitz A, Loke SW, Zaslavsky A (2004) Towards a theory of context spaces. In: Proceedings second IEEE annual conference on pervasive computing and communications. Workshops, PerCom
18. Welty C, Fikes R, Makarios S (2006) A reusable ontology for fluents in OWL. In: Formal ontology in information systems. IOS Press
19. Bucks RS, Ashworth DL, Wilcock GK, Siegfried K (1996) Assessment of activities of daily living in dementia: development of the bristol activities of daily living scale. age ageing
20. Lawton MP, Brody EM (1969) Assessment of older people: Self-maintaining and instrumental activities of daily living. Gerontologist
21. Katz S, Downs TD, Cash HR, Grotz RC (1970) Progress in development of the index of ADL. Gerontologist
22. Bartos A, Martínek P, Řípová D (2010) The bristol activities of daily living scale BADLS-CZ for the evaluation of patients with dementia
23. Bucks RS, Haworth J (2002) Bristol activities of daily living scale: a critical evaluation. Expert Rev Neurother
24. Philipose M, Fishkin KP, Perkowitz M, Patterson DJ, Fox D, Kautz H, Hähnel D (2004) Inferring activities from interactions with objects
25. Riboni D, Pareschi L, Radaelli L, Bettini C (2011) Is ontology-based activity recognition really effective? In: 2011 IEEE international conference on pervasive computing and communications workshops, PERCOM Workshops 2011
26. Horrocks I (2005) OWL: A description logic based ontology language. In: Lecture notes in computer science (including subseries Lecture Notes in Artificial Intelligence and Lecture Notes in Bioinformatics)
27. Mann CJH (2003) The description logic handbook – theory, implementation and applications. Kybernetes
28. Baader F, Calvanese D, McGuinness DL, Nardi D, Patel-Schneider PF (2010) The description logic handbook: theory implementation and applications. Cambridge University Press, New York
29. Horrocks I, Patel-Schneider PF, Bechhofer S, Tsarkov D (2005) OWL rules: A proposal and prototype implementation. Web Semant
30. Bellifemine F, Poggi A, Rimassa G (2001) JADE: a FIPA2000 compliant agent development environment. In: international conference on autonomous agents and multiagent systems
31. Stanford University, University, S.: Protégé
32. Stardog-union: Pellet: OWL 2 Reasoner for Java, https://github.com/stardog-union/pellet
33. Friedman-Hill E (2008) Jess The rule engine for Java Platform

34. Jing Mei, EP Bontas: Technical Reports: Reasoning Paradigms for Owl Ontologies. http://www.ag-nbi.de/research/owltrans/
35. Chan M, Campo E, Estève D, Fourniols JY (2009) Smart homes---Current features and future perspectives

Chapter 8
Semantic Smart Homes: Towards a Knowledge-Rich Smart Environment

8.1 Introduction

Recently the provision of health and social care is undergoing a fundamental shift towards the exploitation of technologies to support independent living. These efforts have been driven by the ever-growing ageing population and the increasingly over-stretched healthcare resources. Smart Homes (SH) have emerged as a mainstream approach to enable the use of technologies in an individual's living environment to facilitate independent living. SH are augmented environments equipped with sensors, actuators and devices, inhabited by the elderly or disabled and monitored/supported by professionals and health services. The primary impetus for SH research and development stems from the personal preferences of people to remain in their own home even if they appreciate that they may be at risk. Additionally, SH are able to support user-centred personalised healthcare, thus offering the potential to enhance the quality of life for people at home.

There are currently a number of SH [1–5] in development for the purposes of demonstration as well as for the establishment of real living environments. Researchers are using a multitude of technologies that can provide individual aspects of the functionality required for SH. For example, technologies in sensor networks, wearable systems, smart devices and Information and Communication Technologies (ICT) have all been developed for the capture, communication and analysis of data pertaining to environments, inhabitants and events within smart homes. In the communication layer, open standards and protocols [6, 7] have been developed to address data exchange and compatibility issues among different types of devices and services. In activity tracking, monitoring and recognition, approaches and technologies have been researched and experimented aiming to capture, re-construct and further advise the behaviour of smart home inhabitants [8–10].

It may, however, be considered that current endeavours in both technologies and solutions are ad hoc, environment dependent and scenario specific. In most cases, data collected from one sensor are used only for one purpose and then discarded. Technological solutions are often developed for well-defined specific cases. It is

L. Chen and C. D. Nugent, *Human Activity Recognition and Behaviour Analysis*, https://doi.org/10.1007/978-3-030-19408-6_8

therefore difficult, if not impossible, for them to be applied in a similar situation, usually requiring substantial re-engineering. At present, large-scale sensory data from sensors, inhabitants, environments and external sources can be captured and collected. Nevertheless, these raw data are too primitive to be processed, used and reused, effectively and intelligently. Though many data processing technologies have been developed, the provisioning and deployment of a generic solution by integrating these fragmented, disjointed technologies is clumsy. And, they are not scalable and feasible in real-world situations.

The reason for the aforementioned problems can be associated with the fact that existing SH technologies and infrastructure are not built upon a commonly agreed SH data model at both data and application levels, together with an expressive representation. This gives rise to three direct consequences in the development and deployment of SH-based solutions:

- Data heterogeneity hinders seamless exchange, integration and reuse of data.
- Application heterogeneity disallows component (i.e., middleware services) reuse in different application scenarios.
- Without the support of formal data models and expressive representation formalisms, current SH technologies are incapable of dealing with rich metadata and their semantics.

The lack of semantics and inability of data sharing, and integration reduce the potential to carry out deep, intelligent data analysis and knowledge discovery from multiple data sources, such as trend discovery, pattern recognition and knowledge-based decision making. This ultimately leads to the difficulty of developing and deploying systematic SH solutions with seamless data integration and advanced high-levels of intelligent capabilities.

As such, there is currently a major gap between these endeavours and the aspiration of what SH should achieve. A vision which can bridge this gap embraces technical solutions with a high degree of easy-to-use and seamless automation along with flexibility and scalability in system reconfiguration and deployment, and with adaptation, personalisation and context-awareness in assistance provisioning. In this chapter, we propose the concept of the Semantic Smart Homes (SSH). This concept may be viewed as going beyond current SH technologies through the creation and management of large-scale, rich semantic information, thus enabling and supporting high-level intelligent capabilities. The cornerstone for the SSH is the ontology-based approach to data modelling for SH entities, including inhabitants, environments, devices and services. Semantic modelling offers realistic solutions to a number of research issues faced by SH based assisted living such as data interoperability, integration and semantic/knowledge-based intelligent decision-making support.

8.2 Semantic Smart Homes

8.2.1 The Concept

We define SSH as an extension of the current SH in which data, devices and services are given well-defined meaning. This will better enable the environment, devices, services/applications and people (inhabitants and professional carers) to work in cooperation through the extraction of more meaning from the data collected and more appropriate support measures offered to the inhabitant. The essence of a SSH is to have data within and across SH defined and linked in a way that it can be used for more effective discovery, processing, automation, integration and reuse across various applications. Specifically, with semantics and relationships in place we can exploit advanced semantic or Artificial Intelligence (AI) based information processing techniques to provide value-added data processing capabilities such as data integration, interoperability and high-level decision support within and across SH communities [11, 12]. We envisage that the SSH notion will bring the semantic dimension into SH solutions, enable semantic-based knowledge discovery and intelligent processing as has been witnessed within the general Semantic Web community. This will allow us to ultimately move from the current state of the art of SH technologies to the next generation SH infrastructure that is required to address current shortcomings.

Central to the SSH concept and its realisation is semantic data modelling and representation. The rationale associated with this concept is that the more semantics and knowledge the data model can hold and represent, the more capabilities and flexibilities that SSH technologies and applications can achieve in the processing of data.

We contend that ontologies and the Semantic Web infrastructure are the enabling technologies for the realisation of SSH. An ontology is an explicit, shared specification of the various conceptualisations in a problem domain. It defines commonly agreed data/knowledge structures, i.e., domain concepts, their attributes and the relations between them. In addition, it also provides a shared vocabulary for describing these structures. This means that data providers, no matter where they are, can use these same structures to preserve and publish semantic-rich data and equally consume data from other sources. Ontologies provide a common medium for inter-agent information exchange, interoperation and integration. As ontologies specify the semantics of terms at the conceptual level based on the explicit conceptualisation of a domain, they are understandable and easily processed by both humans and machines, thus increasing the potential of automation.

8.2.2 Related Work

Ontology-based modelling has been extensively explored in the domain of context modelling. This strand of work concentrates on the modelling of high-level abstract contextual concepts and/or facts, such as person, location, activity and computational entity, either sensed, static or profiled, with constraints and annotations. While it provides some general guidance as a type of upper-level ontology, it fails to capture specific characteristics of SH. Already, there have been several attempts to use ontologies to model context in an assistive living scenario [13, 14]. The use of existing ontologies is primarily restricted to specific pervasive aspects and usually for reasoning purposes only.

Using ontologies for SH modelling can be viewed as a recent development. Latfi et al. [15] proposed an ontological architecture for a Telehealth Smart Home (TSH) and developed prototype ontologies. Klein et al. [16] proposed a context ontology for ambient middleware as part of the European Union funded SOPRANO project. They claimed that ontologies will be used as a central reference document for SOPRANO middleware. Both studies are at concept level. It is not clear how large-scale semantic content is created and used in real-world scenarios.

While our research shares some consensus with [17] in ontology modelling and with [15] in the role and use of ontologies, it is fundamentally different from these works in that we take a broad, integrated and systematic view towards SSH. In this case, ontology-based semantic modelling and representation is not just used for separate, stand-alone components for some specific purposes. Rather, ontologies are regarded as a conceptual backbone and a common vehicle for enabling and supporting communication, interaction, interoperability, integration and reuse among devices, environments, inhabitants and external sources. Our focus is on how to capture and model rich semantic metadata with the emphasis being placed on the effective use and reuse of intelligent content for supporting assistive living. We also address issues pertaining to semantic data lifecycle management, namely modelling, creation, storage, use/reuse and maintenance.

There have been previous projects such as European Union projects ASK-IT [18] and SAPHIRE [19], which intend to use Semantic Web technologies for interoperability and integration. Until now the SSH concept, and in particular, the idea of using ontologies as the conceptual backbone for integration, interoperability and high-level intelligent processing, has not been witnessed.

8.2.3 The Conceptual Architecture

We propose a layered conceptual architecture for the SSH, as shown in Fig. 8.1. The Physical Layer consists of physical hardware such as sensors, actuators, and various devices including medical equipment, household appliances and network components. This layer provides the means to monitor and capture the events and

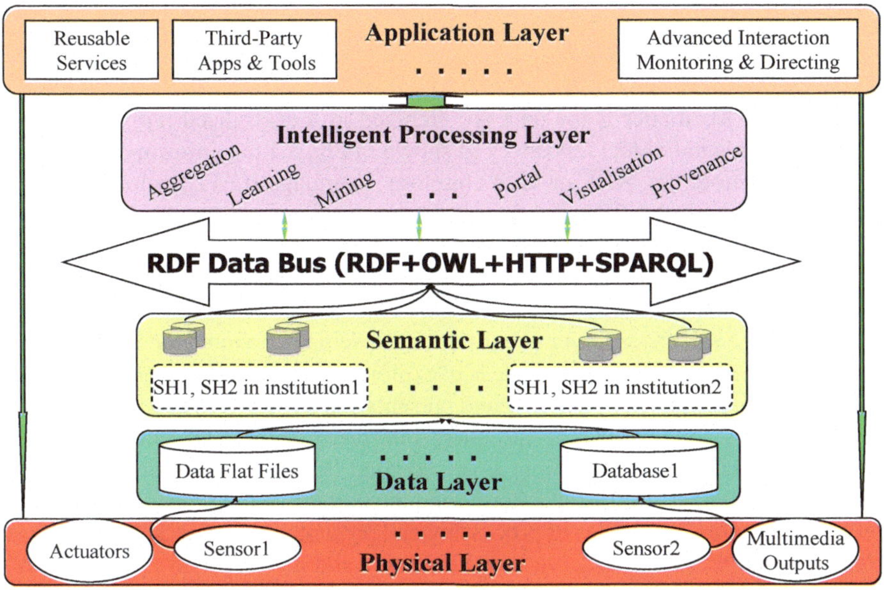

Fig. 8.1 The conceptual architecture of the SSH

actions in a SH, and subsequently traverse data to the Data Layer. The Data Layer archives collected raw data in a number of data stores. These stores are usually disparate in data formats and access interfaces, with each of them being dedicated to individual sensor-based application scenarios. The Application Layer contains various capabilities, tools and (sub)systems for assistive living. Within this layer applications process sensory data which has been passed via the Data Layer and can be used to control actuators and/or multimedia facilities in the Physical Layer to offer assistive living as required. These three layers have so far been the core conceptual components underpinning SH application design and development. While each layer is indispensable for any SH application, the close coupling among sensors, data and applications, often having one to one, ad hoc relationships, causes many challenges as discussed in the introduction.

The SSH addresses these challenges by incorporating a Semantic Layer, a RDF (Resource Description Framework[1]) Data Bus and an Intelligent Service Layer in the proposed systems architecture. These layers break down the direct links between the Data and Application Layers and provide underpinning technologies for data sharing, reuse and application development. The goal of the Semantic Layer is to provide a homogeneous view over heterogeneous data, thus enabling seamless data access, sharing, integration and fusion across multiple organisations. It achieves this by using SH ontologies as a unified conceptual backbone for data modelling

[1] RDF, RDFS, OWL, HTTP, SPARQL and URI that will be mentioned later are all W3C standards. Detailed information can be found at https://www.w3.org.

and representation. Semantic modelling allows the markup of various data with rich metadata and semantics to generate semantic content. Multiple SH in geographically distributed locations supported by various organisations can then aggregate and fuse their SH data. No matter if the data are archived in a centralised repository or in each institution's individual repository as shown in Fig. 8.1 the uniform data models and representation, e.g. RDF or Web Ontology Language (OWL), allow seamless data access through the RDF Bus based on the standard communication protocol HTTP and RDF query language SPARQL. The Semantic Layer is also responsible for providing tools and APIs for semantic data retrieval and reasoning. Details will be presented in next section.

The Intelligent Service Layer is built upon the semantic content and functionalities of the Semantic Layer. Its purpose is to exploit semantics and descriptive knowledge to provide advanced processing and presentation capabilities and services. The former provides added-values to the query interfaces of the RDF Bus, by further searching, analysing and reasoning over recorded SH data. The latter essentially visualises the contents of the repositories and the outputs of the processing services. Table 8.1. lists some examples of processing and presentation services. Such a list of processing and presentation services is illustrative and not exhaustive; furthermore, it does not mean that each SH will use all these services. In fact, the selection and use of such services will depend on the nature and availability of collected data as well as the personal needs of inhabitants and care providers, hence exploiting further the concepts of personalisation. These services can be realised using industry standards such as Web services [20] and are given well-defined meaning, e.g. semantic Web services [21]. They are accessible to third-party developers, thus interoperable and reusable at both the service and application level.

Table 8.1 A list of examples processing and presentation services

Processing services	Presentation services
Compare activities of daily living (ADL) of subjects in the same group and/or different groups	Browsing and navigation facilities over a single or federated SH repositories
Aggregate multiple data sources to create a single virtual large data set for data mining	Visualise ADL and their differences of subjects in the same group and/or different groups
Offer semantic based search and discovery	Illustrate relationships of ADL from a more semantic viewpoint
Extract user profiles and ADL patterns	Present graphically the results of the various statistical and probabilistic analysis in mining and learning
Create inhabitant communities for social or medical purposes	Allow professional carers to specify inhabitants' ADLs in a graphical manner
Re-construct an ADL trace from a living context	

The Semantic Layer essentially achieves data interoperability and machine understandability, whereas the Intelligent Service Layer delivers the capability of interoperability and high-level automation. As such, the proposed architecture enables a novel and flexible paradigm of SH system development and deployment. In this paradigm services in the Intelligent Service Layer are responsible for data access and the provision of processing and presentation capabilities. They have well-defined interfaces and can be boxed as primitive off-of-shelf building blocks. SH systems shall have little direct interaction with data at the lower layers. SH system development will be accomplished by the aggregation and assembling of various on-demand services in terms of SH system requirements. New functionalities, i.e. capabilities and services, can be made available whenever needed. Eventually, a robust feature-rich technological infrastructure will be in place to facilitate the delivery and agile deployment of assistive living solutions, e.g. plug and play and open system development.

8.3 Semantic Smart Home Analysis

A SH is a complex ecosystem typically consisting of a physical environment with various furniture, household appliance and rooms, one or more inhabitants that perform various ADLs within the environment, and sensors and devices (actuators) to sense and act on environmental changes and inhabitant behaviours. At any specific time, it will generate data/information about the environment such as temperature, humidity, the status of doors, windows and lights, about the behaviours of inhabitants such as sleeping, cooking or watching TV and about events within the smart home such as alarm-fired, cooker-turn-on or tap-turn-on. Such information once monitored and collected can be aggregated to denote a situation against which an assistive system should be able to carry out interpretation and reasoning to make just-in-time assistance for the inhabitant. As such, the central issue is therefore how to aggregate data from multiple data sources to form a meaningful situation and further interpret them at a higher level of automation, i.e., by software agents.

The nature of SH presents a number of challenges to situation formation and comprehension. Firstly, most sensor data are primitive numerical data such as 3-D coordinates for motion detectors, 2-state values for contact sensors. They lack formal descriptions and can only be consumed by humans through hard-coded operation logics in ad hoc data processing components. It is difficult, if not impossible, for machines or soft-agents to interpret and reason their high-level situational meaning. Secondly, sensor data are increasingly available in a variety of diverse forms, such as unstructured textual data, audio and surveillance videos. They are heterogeneous in data formats and representation, and conceptually isolated from each other. For example, a location sensor (or a video monitor) can detect an inhabitant in front of the cooker. An event sensor detects the turn-on of a cooker and a contact sensor detects the movement of a spaghetti pack. While each sensory data reflects one facet of the situation, it requires the interlinking and fusion of data from multiple, disparate information sources in order to comprehend and understand such a complex situation.

In addition to perceived data, the third challenge is how to model and represent normal routine ADLs and an inhabitant's profiles. ADLs are actually the common-sense knowledge and heuristics of daily living, which provides interrelationships among environments, events and activities within an SH. For instance, cooking takes place in the kitchen using a cooker around three times a day. Making a cup of tea involves the preparation of hot water, tea bag, milk and/or sugar. User profiles, on the other hand, describe personal information including an inhabitant's personal medical data, hobbies, favourite activities and activity patterns. User profiles can be used to provide inhabitant dependent personalised assistance. Formal modelling and representation of ADLs and user profiles, in essence, provide a recognition context for an assistive system to interpret perceived situational data for the provision of personalised assistance.

8.3.1 Semantics, Semantic Modelling and Representation

Semantics refers to meaning. Semantic modelling refers to the process of defining the meaning of data, devices and services. The basic formalism used for semantic modelling is the RDF. RDF is a graph data model for describing resources and relations between them. An RDF graph contains a set of triples, each triple consisting of the subject, predicate (property) and object. This structure can be considered as a natural way to describe the vast majority of the data processed by machines. A triple can make assertions that particular things (such as sensors, inhabitants) have properties (such as "is used for", "has a type") with certain values (thermostat, dementia). Figure 8.2 shows the RDF graph that represents the group of statements "there is a Contact Sensor identified by http://www.ulster.ac.uk/ssh2008/ssh#Trail_lab_contact_sensor_9, and it *is located in* the second *cupboard kitchen_cupoard_2* of the kitchen, attached to the bottle *milkBottle* of milk and it activates the *milk_moved* event.

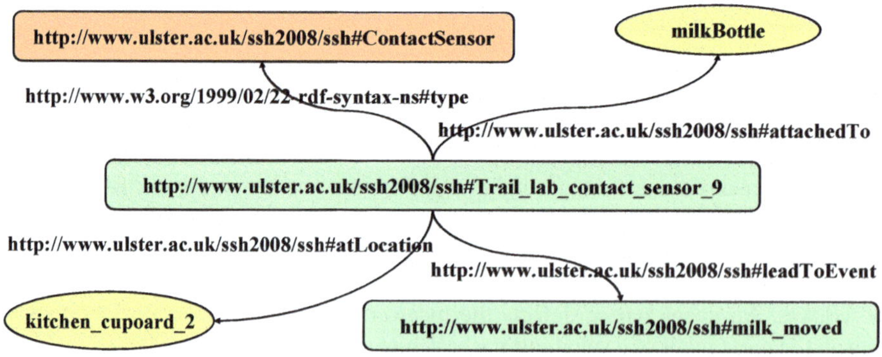

Fig. 8.2 An RDF graph describing a concrete contact sensor

RDF encodes meaning in their triple statements. A single separate data item does not stand for anything. Its meaning can only be interpreted against a context in which the term appears. In an RDF expression, the subject, predicate and object are all identified by a Universal Resource Identifier (URI)—see Table 8.2. This ensures that concepts and properties are not just terms (keywords) in a domain but can be tied to a context where their unique definitions can be interpreted. The context is in essence, the ontologies that formally define all core concepts and the relations between them. A typical ontology usually contains a hierarchical structure of concepts and subconcepts. Relations between concepts are established by assigning properties to concepts and allowing subconcepts to inherit such properties from their parent concepts. In the above example, the ontology is defined in http://www.ulster.ac.uk/ssh2008/ssh.

Ontology languages such as the RDF Schema (RDFS) and OWL are used to specify domain concepts and relationships between these concepts. RDFS defines a vocabulary (terms) for describing the properties and classes of RDF resources, with semantics for generalisation hierarchies of such properties and classes. Table 8.2 is the RDF representation of the RDF graph in Fig. 8.2. On top of RDFS OWL adds more vocabulary for describing properties and classes: among others, relations between classes (e.g. disjointness), cardinality (e.g. "exactly one"), equality, richer typing of properties, characteristics of properties (e.g. symmetry), and enumerated classes. This gives OWL more expressive power for representing complex data semantics.

Ontology languages carry built-in inference rules from underlying data models such as RDF graphs or OWL's Description Logics. This subsequently gives semantic representation further power to enable inference and reasoning via the notion of entailments. An ontology may express the rule "If an inhabitant's action is in response to an event, and a sensor generates that event from one change, then the inhabitant's action can be associated with that change". A program could then deduce, for instance, that the action is the direct reaction to the change. The computer doesn't truly "understand" any of this information, however, it can now manipulate the terms much more effectively in ways that are useful and meaningful to human users.

8.3.2 Smart Home Ontology Engineering

Ontology development is a formal process of knowledge acquisition and modelling. It requires the close cooperation of domain experts and knowledge workers.

Table 8.2 The RDF representation of the RDF graph in Fig. 8.2

```
<ContactSensor rdf:ID="Trail_lab_contact_sensor_9">
      <leadToEvent rdf:resource="#milk_moved"/>
      <attachedTo
      rdf:datatype="http://www.w3.org/2001/XMLSchema#string">milkBottle
       </attachedTo>
      <atLocation rdf:resource="#kitchen_cupoard_2"/>
</ContactSensor>
```

Domain experts identify and describe concepts, properties, their relations, instances and role-playing actors within a problem domain, and domain-specific, application-dependent problem-solving processes. Knowledge workers, who do not have domain background, will use extensive knowledge engineering techniques to capture useful knowledge based on the experts' expertise and will develop knowledge-preserving structures, i.e. models, which can hold and share reusable information. As such, domain analysis and characterisation is essentially the first step of ontology development.

A SH is a complex micro-ecosystem that usually consists of the following constituents:

- a physical environment with various pieces of furniture, electrical/electronic household appliances, and rooms which provide a living space,
- inhabitants that perform various activities within the environment,
- sensors, actuators and medical devices to sense and act on environmental changes and inhabitant behaviours,
- assistive resources including actors (care-providers or family members), middleware services or applications to respond to events and situations.

Each of these constituents plays an indispensable role and provides specific functions. Overall, they deliver 'just-in-time' assistance for inhabitants through intercommunication and causal interactions.

Based on the above characterisation a SH can be modelled in seven ontologies. They include an ontology for the physical equipment such as sensors, actuators, medical devices and home electronic or electrical appliances; an ontology for actions and ADLs such as watching television and making drinks; an ontology for living spaces and environments such as the kitchen, sitting rooms; an ontology for actors such as inhabitants, care-providers; an ontology for medical information; an ontology for software components such as services and applications and an ontology for time in order to model temporal information. Each ontology is used to explicitly conceptualise a specific aspect and overall, they provide a semantic model for smart homes. Details of semantic modelling, semantic data management and activity recognition are described in Chap. 9. It is worth noting that existing well-defined ontologies could be imported and reused directly, for example some medical ontologies and a time-based ontology [22].

8.4 Semantic Enabled Processing Capabilities

Semantic modelling gives data many characteristics that are otherwise not available. Firstly, it enables the data to be exchangeable, interoperable and accessible at both intra- and inter-institutional levels based on the commonly accepted ontological schema. Secondly, it makes data understandable and easily processed by both humans and machines (or software agents). Thirdly, semantic data supports reasoning and inference by incorporating entailment rules in expressive representation. These attributes make semantic data amenable for flexible and complex manipulation, thus enabling many advanced processing capabilities such as automated processing and knowledge discovery, and novel application scenarios such as data sharing, reuse, integration, and situation-aware assistance.

Given the manner in which semantic data are used is only limited by the application's requirements and the developer's imagination, it is unwise and practically impossible for us to try to elaborate all usage mechanisms. As such this section will omit discussions relating to the basic use of semantic data, for example how to facilitate data sharing and exchange, how to carry out semantic retrieval and searching, as these features have already been elaborated and illustrated through research results in various domains. Instead, we shall discuss some core innovations brought specifically to SH through the combination and synergy of these semantic data properties.

8.4.1 Towards a Paradigm of Extensible and Flexible Assistance Provisioning

The SSH concept can support an open, plug-and-play paradigm that makes assistance provisioning extensible and flexible, and facilitates rapid prototyping and is easy-to-deploy. This paradigm is enabled by the explicit separation and management of entities, i.e. devices, inhabitants and individual SH, and functionalities, i.e. services and applications. It distinguishes services, i.e. high-level functional components that are used as building blocks for multiple applications, from applications, i.e. systems that are used by end users, either care providers or inhabitants for providing assistance. In particular, it unties the direct links between services and applications and specific devices, environments (individual SH) and inhabitants.

As shown in Fig. 8.3, entities in all smart homes within an institution and services (i.e. functions) that are applicable to various data are semantically described and placed in a registry. Care providers in the Central Assistance Provisioning Environment in an Institution can discover applications from the registry based on requirements of individual inhabitants. The discovered applications are then linked to sensory data and provide assistance through data processing. In this paradigm, new entities and functions, such as a newly installed sensor, a new resident or a new function, can then be added into the registry anytime for discovery and reuse while the whole system is still working. Therefore, it supports the plug-and-play concept and makes the system extensible. Application developers can discover and reuse available services to develop new applications, and can then publish the application in the registry for further reuse. This paradigm will significantly reduce the needs for developing new services and applications when new SH or devices are added. Equally services and applications requiring sensory data can search the registry to discover available devices that provide the required data. This paradigm saves not only effort for the development and cost for new devices, but, also facilitates rapid prototyping and easy deployment.

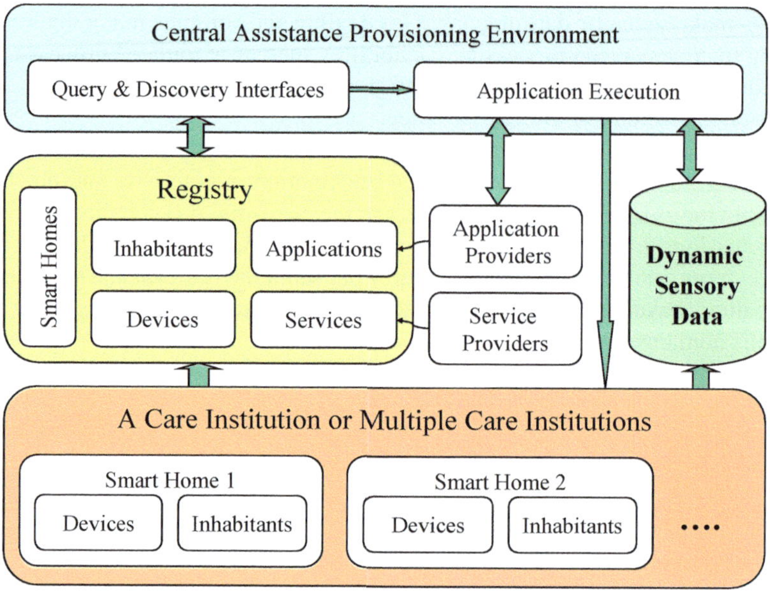

Fig. 8.3 An open paradigm for assistance provision

The above discussion is made in the context of an SH based care institution(s). It is actually applicable to more generic scenarios. For example, individual SH could be geographically dispersed across a wide area without belonging to any specific care institutions. In an extreme case, ordinary family homes could be connected to such assistive systems through broadband, passing data and receiving advice. The key point we wish to make is that the SSH concept enables an open home paradigm for assistance provision.

8.4.2 Cognitive ADL Monitoring and Recognition

Current SH can provide reminding assistance through pre-defined instructions, such as the instruction to take medicines as a specific time [23], to perform ADLs such as prepare a meal [24] and reactive emergency handling, for example, calling for a fire engine when a fire breaks out. In the future, however, it is increasingly expected that assistance at a behavioural level is provided for the elderly. A particular group within this cohort who would benefit from the deployment of SH technology would be those suffering from cognitive deficiencies such as Alzheimer's disease. For these persons, it then becomes necessary to monitor their behaviour and recognise their intended ADLs so that just-in-time assistance can be provided.

Semantic modelling and reasoning can achieve this in a scalable and automatic way by building ontological behavioural models. The basic idea is that through semantic modelling we can build an ADL ontology as shown in Fig. 8.4 with each node denoting a type of ADL. Each ADL class is described with a number of properties and sub-classes can inherit all properties from its parent class. A property is defined by specifying its domain and range. The domain refers to all classes that can be described by the property and the range refers to all classes whose instances can be assigned to the property. A property describes a class using either a literal or an instance of another class as its value, thus linking two classes. This essentially gives rise to a description based behavioural model, i.e., an ADL is described by various properties.al model, i.e., an ADL is described by various properties.

The underlying mechanisms for ADL monitoring and recognition are straightforward and natural. If we can identify a number of properties, then we could infer and recognise an ADL or ADLs based on the described ontological behavioural model. In semantic modelling, the perception of an event and/or the detection of sensory signals imply the identification of a concrete instance of a class. For example, the activation of a contact sensor in a cup means that the cup, as an instance of Container, is used in an ADL. As the Container class is the range of the *hasContainer* property, it can be inferred that the *hasContainer* property is assigned the value cup. Since the *hasContainer* property is used to describe the *MakeDrink* class, it can be further inferred that a *MakeDrink* ADL has taken place. Nevertheless, it is not possible to ascertain whether the ADL is *MakeHotDrink* or *MakeColdDrink* as both ADLs have the *hasContainer* property. This is exactly one of the advantages of the description-based ADL recognition because based on limited sensory information the system can still identify uncertain high-level ADLs. In the given example, though we cannot tell the concrete ADL, i.e. the *MakeHotDrink* or the *MakeColdDrink*, we can at least know that the inhabitant is performing a *MakeDrink* ADL. When more sensory becomes available, concrete ADL(s) can be identified. Suppose that a contact sensor in a tea container is activated, this implies that an instance of the *HotDrinkType* class,

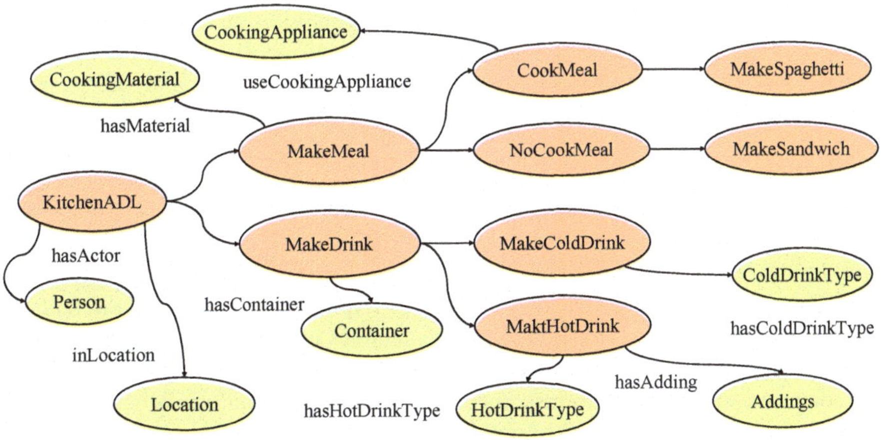

Fig. 8.4 A fragment of the graphical hierarchy of the ADL ontology

i.e. the tea, has been specified, and the *hasHotDrinkType* is assigned the value tea. In this case, it is reasonable to assume that an ADL or ADLs that happen, though we do not know yet, must have at least the two properties *hasContainer* and *hasHotDrinkType*. Based on the ontological ADL model we can infer that it is the *MakeHotDrink* ADL.

The ADL monitoring and recognition process can be summarised as follows: Sensory inputs are used to identify concrete items that have already been specified as instances of classes in SH ontologies. In terms of the scope of a property's range, the property that takes the identified item as its value can be inferred. In terms of the scope of a property's domain the ADL(s) that can be described by the inferred properties can then be recognised. As properties can be inherited from super-classes (higher level abstract ADLs) to sub-classes (lower level concrete ADLs), the lower a class is in the ADL class tree the more properties it has. This means that the more sensory data that are available, the more accurately ADLs can be recognised. For example, if we only have the location sensory data as *inKitchen*, we can only infer the inhabitant might perform a *KitchenADL* at that specific point in time, without knowing what ADL it is. If further data become available, for example cup sensory data, then we can infer the inhabitant might perform the *MakeDrink* ADL at that specific point in time. Nevertheless, we still do not know what drink the inhabitant will make. If we obtain the coffee sensory data, then we can determine that the inhabitant is making coffee, but we still do not know if it is a white coffee or a black coffee. Hence the sensory data from milk or sugar sensors can further help to recognise the details of the performed ADL. From what we have described above, it is apparent that the proposed approach can monitor the unfolding of an ADL and incrementally recognise the ultimate ADL, which may be considered as not previously possible. The monitoring and recognition process conceptually corresponds to the subsumption operation of description logic based reasoning, which can be realised using reasoners such as FaCT or RACER [25].

The semantic model which enables behaviour monitoring and recognition has a number of compelling advantages: Firstly, the scalability of SH ADL modelling has been a bottleneck to effective behavioural recognition. It is often the case that proof-of-concept experiments, either state-based or process-based approaches, work well but fail to scale up. Ontological ADL modelling does not have this problem with the extensive technological support which is offered in ontology engineering, which includes tools, APIs, storage and reasoners. Ontologies of thousands of classes have been developed in other domains, e.g., 7,000 concepts in the gene ontology, and semantic data repository of 25 million triples is also practiced in TripleStore [26]. For smart homes, ADL classes and associated instances are simply not present in such scale. Secondly, semantic ADL models contain explicit rich semantics and built-in logical entailment rules. This allows not only humans but also software agents (such as assistive systems) to interpret, comprehend and reason against captured semantic ADL data. As such, behaviour monitoring and recognition can be realised at higher levels of automation. Thirdly, description-based reasoning provides a mechanism to incrementally predict ADLs by interpreting limited or incomplete sensor data. This capability is particularly important because assistive systems are supposed to provide reminding or suggestive assistances with limited sensory data.

8.4.3 Knowledge-Based Assistive Living Systems

Ontologies are knowledge models. ADL classes and their hierarchical structure in SSH ontologies are in essence the explicit model and representation of the common-sense knowledge of a human's daily activities and their classification. In addition to these generic ADLs, individual inhabitants have their own living habits, regular ADL patterns, preferences and unique ways to respond to various events. Such individual lifestyles may further vary with age profile, culture and personality. Using semantic modelling we can formally capture, model and represent an individual inhabitant's personal specialties in semantic repositories. These heuristics and knowledge can then be exploited for intelligent living assistance. A typical example is personalised assistance. Consider that a *MakeDrink* ADL is recognised as described in Sect. 9.3 for a person with dementia. An example of general assistance provided would be to advise the person to make either a cold drink or a hot drink. Then the assistive system will monitor the person's behaviour and advise its actions accordingly. If the person's preferences on making a drink are known, e.g., she/he likes hot white coffee, then the assistive system can directly remind the person what she/he should do in order to make their coffee hot with milk. Similar assistance can be offered for recommending other ADLs, for example TV channels, etc.

Another knowledge-based use scenario is adaptive assistance. Rather than modelling an inhabitant's behavioural preferences a priori, an assistive system can derive an inhabitant's ADL patterns through data mining and pattern recognition against collected semantic data. This will capture the evolution of an inhabitant's daily life and incorporate changes into behavioural models. An assistive system can then reason against learnt ADL patterns to provide adaptive assistance.

8.5 Summary

Research on SH and assistive living has come to a critical point where novel paradigms and technologies are needed in order for the approach to be useful in real-world scenarios in terms of applicability, scalability and ease of use. This chapter has introduced the concepts of SSH that aims to break down barriers (heterogeneity) and isolations (hardwired) among devices, data, capabilities and applications, and to unleash the potential of the approach through semantics, rules and expressive representation. We have proposed an integrated systems architecture for SSH and discussed its core functional components and interplay. We have described in detail the methodology and related technologies for semantic modelling and semantic content management.

While the SSH concept, its enabled assisted living paradigm and underpinning technologies await further investigation, development and evaluation through real world use case studies, the work presented in this chapter has laid a solid architectural and methodological foundation. Initial results have demonstrated the potential and value of the approach and further clarify future research directions. We believe that SSH are the next generation of technological infrastructures for assisted living that facilitates the innovative exploitation of research results from AI, Web technologies and information processing.

References

1. Rochester University: University Creates Medical "Smart Home" To Study Future Health Technology - Newsroom - University of Rochester Medical Center. https://www.urmc.rochester.edu/news/story/-103/university-creates-medical-smart-home-to-study-future-health-technology.aspx
2. Nugent CD, Mulvenna MD, Hong X, Devlin S (2009) Experiences in the development of a smart lab. Int J Biomed Eng Technol 2:319–331
3. MIT: House_n The PlaceLab. http://web.mit.edu/cron/group/house_n/placelab.html
4. Georgia Institute of Technology: Aware Home Research Initiative. http://www.awarehome.gatech.edu/
5. Espinilla M, Martinez L, Medina J, Nugent C (2018) The experience of developing the UJAmI Smart Lab. IEEE Access 6:34631–34642
6. OSGi Alliance: OSGi™ Alliance – The Dynamic Module System for Java. https://www.osgi.org/
7. KNX: KNX Internet of things KNX Association. https://www.knx.org/knx-en/for-professionals/benefits/knx-internet-of-things/index.php
8. Salguero AG (2018) Using ontologies for the online recognition of activities of daily living. Sensors (Basel) 18:1–22
9. Bibi S, Anjum N, Sher M (2018) Automated multi-feature human interaction recognition in complex environment. Comput Ind 99:282–293
10. Almeida A, Azkune G (2018) Predicting human behaviour with recurrent neural networks. Appl Sci 8:305
11. Davies J, Studer R, Warren P (2006) Semantic Web technologies: trends and research in ontology-based systems. Wiley, New York
12. Pollack ME, Pollack ME (2005) Intelligent technology for an aging population: the use of AI to assist elders with cognitive impairment. AI Mag 26(2):9
13. Noor MHM, Salcic Z, Wang KIK (2018) Ontology-based sensor fusion activity recognition. J Ambient Intell Humaniz Comput 1–15 (2018)
14. Meditskos G, Kompatsiaris I (2017) iKnow: ontology-driven situational awareness for the recognition of activities of daily living. Pervasive Mob Comput 40:17–41
15. Latfi F, Lefebvre B, Descheneaux C (2007) Ontology-based management of the telehealth smart home, dedicated to elderly in loss of cognitive autonomy. In: CEUR workshop proceedings
16. Klein M, Schmidt A, Lauer R (2007) Ontology-centred design of an ambient middleware for assisted living: the case of SOPRANO. In: Towards ambient intelligence: methods for cooperating ensembles in ubiquitous environments (AIM-CU), 30th annual German conference on artificial intelligence
17. Roussaki I, Strimpakou M, Pils C, Kalatzis N, Anagnostou M (2006) Hybrid context modeling: a location-based scheme using ontologies. In: Proceedings - fourth annual IEEE international conference on pervasive computing and communications workshops, PerCom workshops 2006

18. European Commission - CORDIS: ASK-IT - Ambient intelligence system of agents for knowledge-based and integrated services for mobility impaired users. https://cordis.europa.eu/project/rcn/72134/factsheet/en
19. European Commission - CORDIS: Final Report Summary - SAPPHIRE (System Automation of PEMFCs with Prognostics and Health management for Improved Reliability and Economy). https://cordis.europa.eu/project/rcn/108481/reporting/en
20. W3C: Web Services Description Language (WSDL) Version 2.0 Part 1: Core Language. https://www.w3.org/TR/wsdl20/
21. Johnston WE (2004) Semantic services for grid-based, large-scale science. IEEE Intell Syst 19:34–39
22. The time ontology in OWL. https://www.w3.org/TR/owl-time/
23. Nugent C, Finlay D, Davies R (2007) The next generation of mobile medication management solutions. Int J Electron Healthc 3(1):7–31
24. Philipose M, Fishkin KP, Perkowitz M, Patterson DJ, Fox D, Kautz H, Hähnel D (2004) Inferring activities from interactions with objects. IEEE Pervasive Comput 1(4):50–57
25. Dentler K, Cornet R, Ten Teije A, De Keizer N (2011) Comparison of reasoners for large ontologies in the OWL 2 EL profile. Semant Web 2:71–87
26. Harris S, Gibbins N (2003) 3store: efficient bulk RDF storage. In: Proceedings of the 1st international workshop on practical and scalable semantic systems (PSSS'03)

Chapter 9
Semantic Smart Homes: Situation-Aware Assisted Living

9.1 Introduction

With the advance and prevalence of low-cost low-power sensors, computing devices and wireless communication networks, pervasive computing has evolved from a vision to a realistically achievable and deployable computing paradigm. Research is now being conducted in all related areas, ranging from low-level data collection, intermediate-level information processing, to high-level applications and service delivery. It is becoming increasingly evident that the prevalence of intelligent environments to work and to live in within which flexible multimodal interactions, proactive service provisioning, and situation aware personalized activity assistance, will be commonplace.

At the moment the provision of healthcare is undergoing a fundamental shift towards the exploitation of technologies in pervasive computing to support independent living, as the ever-growing ageing population increasingly over-stretches limited healthcare resources. SH has emerged as one of the mainstream approaches to providing ADL assistances for the elderly, in particular those suffering from cognitive deficiencies such as Alzheimer's disease [1, 2]. A SH is an augmented environment equipped with sensors, actuators, devices and information processing components, inhabited by the elderly or disabled. The rationale is that assisted systems can monitor, collect and process environmental events and user's behaviour through sensors, and respond through actuators or health services, e.g., audio/video outputs or care professionals, to advise the inhabitant the most suitable actions based on the dynamic situation and the inhabitant's ADL profiles.

SH generate massive amounts of data from sensors and mobile devices around the people and entities. Existing research has concentrated on sensor networks, data collection and communication, and can provide low-level ad hoc responsive assistances based on the simple processing of low-level raw sensor data. For example, if a room temperature is lower than a specific value, the air conditioner will start. However, it still remains a challenge to provide just-in-time behavioural and cognitive assistance for cognitively deficient inhabitants such as dementia patients who

L. Chen and C. D. Nugent, *Human Activity Recognition and Behaviour Analysis*, https://doi.org/10.1007/978-3-030-19408-6_9

often get lost during their ADL due to bad memory and/or cognitive problems. For instance, to remind a dementia patient to add milk to a cup after a tea bag and hot water have been added. To achieve this, assistive systems have to be able to observe, interpret and reason the dynamic situations in a SH, both temporally and spatially. In other words, assistive systems should have cognitive capabilities to compensate the loss of the inhabitants' cognition capabilities and to guide the inhabitant's behaviour as normal care providers can do. This further requires that the situational data of a smart home be interpretable and processable by assistive systems.

9.2 Related Work

Making computer systems adaptable to the changes of their operating environments has been previously researched in the context of agent technologies [3]. An intelligent agent is a software system operating in an environment. It senses the changes of the environment, makes a plan in terms of its goal and domain knowledge and takes actions accordingly. An intelligent agent can respond to changes of the environment it inhabits in a number of ways, notably reactive, proactive and adaptive.

Recently technology advances in pervasive computing and ambient intelligence have provoked considerable interest in context-aware applications [4, 5]. Context awareness in pervasive computing refers to a general class of software systems that can sense their physical environments, i.e., their context of use, and adapt their behaviour accordingly. Here contextual information mainly consists of location, time, the entities the system interacts with and the surrounding events and resources. However, context awareness and situation awareness have different research focuses. The former is mainly concerned with linking changes in the environments with software systems. The latter rather concentrates on the knowledge and understanding of the environment that is critical to decision making. Situation awareness pays particular attention on the mental model and cognitive processes from the system's perspective.

Some recent and ongoing work on context aware assistive technologies has adopted an ontology based approach [5, 6]. Nevertheless, ontologies are primarily treated as data models for data/service integration, exchange and sharing in these practices. In contrast, our work uses ontologies as conceptual level knowledge models to support automated situational data interpretation and reasoning.

The use of semantic technologies for situation awareness has been studied in military operational context [7, 8]. While our research shares consensus with these endeavours in using ontologies as the situational data models, the fundamental difference is on how such semantic situational data are used. They have concentrated on semantically enabled data fusion and retrieval. Our work focuses on the innovative exploitation of semantic situational data for the provision of high-level cognitive capabilities with the purpose of delivering cognitive assistance for SH patients. As such we have introduced an agent-based approach to automated situational data comprehension and reasoning. The synergy of semantically enhanced situation awareness with intelligent agents for cognitive ADL assistance has not been seen so far in related research communities.

9.3 A Systematic Approach to Situation-Aware ADL Assistance

A situation is often conceptualized as a snapshot of states at a specific time point in a physical or conceptual environment. Situation awareness has commonly been referred to as "the perception of elements in the environment within a volume of time and space, the comprehension of their meaning, and the projection of their status in the near future" [9]. From this definition we can figure out that situation awareness is a cognitive process that consists of three operational functions. Firstly, it involves the sensing and recognition of different elements in the environment as well as their characteristics and behaviours. Secondly, it needs the interpretation and comprehension of the significance associated with perceived elements in the environment. And thirdly it requires the ability to anticipate the actions of elements and predict future states of the environment. For entities, either human beings or robots or software systems, operating in complex, dynamic and uncertain environments, situation awareness is the determinant of making informed right decisions at the right time in the right place.

Human beings with normal cognitive capabilities are situation aware when they make decisions in their ADL. Nevertheless, SH inhabitants, esp. those suffering from cognitive deficiencies such as Alzheimer's disease, are incapable of doing this. As such a basic requirement of assistive systems is that they should be situation aware. Current SH infrastructure has provided sensor networks for perception, but the interpretation and understanding of perceived data and the realization of high-level cognitive capabilities such as prediction, explanation and planning are still missing.

We propose a systematic approach to enhanced situation awareness for assistive systems, as shown in Fig. 9.1. The approach is grounded on three technological pillars, corresponding to the realization of the three operational functions for situation awareness respectively. The first technological underpinning is based on sensor, device and actuator networks that are responsible for monitoring and collecting situational data. They are mainly embedded in a SH physical environment—see the left Smart Home component in Fig. 9.1. The second pillar is semantic data management as shown in the Semantic Management component, which includes sensor data, ADLs and an inhabitant's ADL profiles. The use of ontologies for data modelling and representation serves two purposes: Firstly, it provides a formal way to model and represent interrelations between data from multiple sources, thus facilitating data fusion and construction of situations. Secondly, it gives data rich metadata and well-defined meaning, thus enabling automated comprehension of the significance of situational data. The third technological pillar is intelligent Assistive Agent that pro-

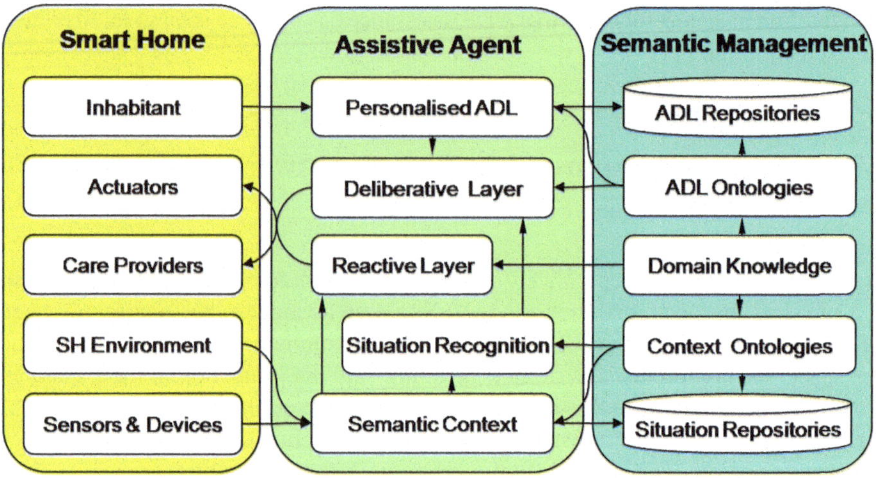

Fig. 9.1 The proposed system architecture

vides high level cognitive capabilities such as prediction, explanation and planning based on reasoning and manipulation of semantic situational data and knowledge. Given the considerable existing work on the physical aspects of SH such as sensors and underlying communication networks, we focus on semantic data management and assistive agent for the realization of situation-aware ADL assistance, which are described in detail below.

9.4 Semantic Data Management

This section describes semantic modelling, semantic content creation and manipulation—the key enabler for the proposed approach. Figure 9.2 depicts the core components and technologies in which ontologies are used as commonly agreed uniform data models to imbue raw data from various data sources with rich metadata and semantics. Both ontologies and generated semantic content are represented using expressive Web ontology languages such as RDF or OWL and are stored in data repositories in which all data are semantically interlinked. Semantic content can be understood and processed by machines or agents, thus allowing a high level of automation, seamless data access, retrieval and reasoning. Details are described below.

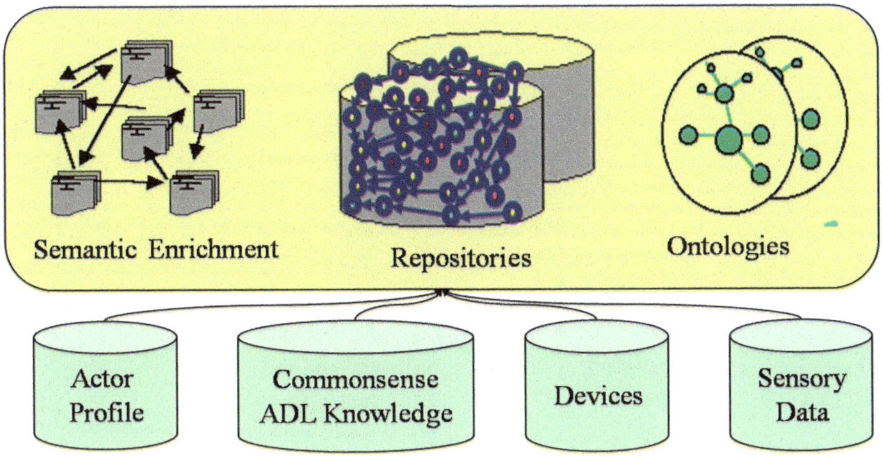

Fig. 9.2 The core components for semantic management

9.4.1 Semantic Data Modelling

Semantic technologies provide a potential solution to the aforementioned challenges. Firstly, ontologies could provide a mechanism for making the meaning of perceived sensor data explicit. This can be achieved through semantic enrichment to create semantic content. Such machine understandable and processable situational data allow assistive systems for automated comprehension of situations. Secondly, semantic modelling and representation facilitate semantically enabled data integration and fusion because situation awareness of complex dynamic environments like SHs often require to fuse information from multiple, disparate information sources for the recognition of a situation [10]. Thirdly, the embedded knowledge such as activity patterns, heuristics and causal relations in ontologies allow assistive systems to reason over perceived situational data with respect to the prediction of future states of SHs or next action of the inhabitant.

Based on the characterisation in Fig. 9.2 we model a SH in seven ontologies. These include an ontology for the physical equipment such as sensors, actuators, medical devices and home electronic or electrical appliances; an ontology for actions and ADLs such as watching television and making drinks; an ontology for living spaces and environments such as the kitchen, sitting rooms; an ontology for actors such as inhabitants, care-providers; an ontology for medical information; an ontology for software components such as services and applications and an ontology for time in order to model temporal information. Each ontology is used to explicitly conceptualise a specific aspect and overall, they provide a semantic data model for the construction of SH situations. Figure 9.3 shows some classes and properties of SH ontologies which have been developed using the Protégé tool. It is worth noting that

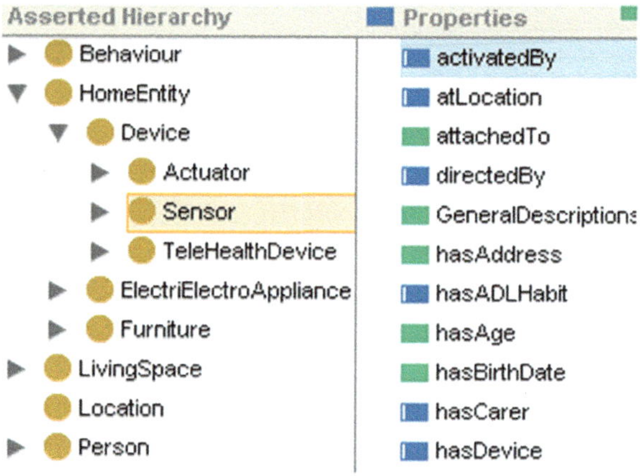

Fig. 9.3 A fragment of the SH ontology

existing well-defined ontologies could be imported and reused directly, for example the time ontology [11].

9.4.2 Semantic Data Creation

Ontologies are knowledge models that can be used to create semantic data. There are two major approaches for this purpose. One is to use generic ontology editing tools such as the Protégé OWL Plugin [12]. These tools can usually be used to perform several activities in one go, such as knowledge acquisition, ontology editing, knowledge population as well as knowledge base creation. They are feature rich but require professional knowledge engineering expertise. So this method is suitable for knowledge engineers. Another approach is to develop domain specific dedicated lightweight annotation tools for domain experts or resource (data) providers to carry out semantic annotation and create knowledge repositories. Such tools are often designed to provide intelligent semi(automatic) support for knowledge acquisition and modelling, including automated information extraction, classification and completion, to help create instances.

Given the nature of data in SH we propose a two phase semi-automatic approach to semantic descriptions. In the first phase data sources such as sensors and devices are manually semantically described. As the number of data sources in a SH is relatively limited, though large, it is manageable to create all semantic instances manually by generic ontology editors such as the Protégé OWL Plugin. Figure 9.4 shows an instance of SSH inhabitant that is created in the Protege and represented in OWL. In the second phase dynamically collected sensory data are first converted to textual descriptors. For example, a contact sensor returns a two-state binary value. It can be pre-processed to literals sensible for denoting two states such as on/off or open/close or used/unused, etc. The concrete interpretation of the state depends on the purpose

```
<Inhabitant rdf:ID="UU_Trail_Occupier">
 <hasFavouriteADL rdf:resource="#WatchUEFAFinal"/>
  <hasPersonalDetail>
      <PersonalDetail rdf:ID="Jemma">
        <hasGender rdf:datatype="http://www.w3.org/2001/XMLSchema#string">Female
        </hasGender>
        <hasTelephone rdf:datatype="http://www.w3.org/2001/XMLSchema#string">
         028 90366666</hasTelephone>
        <hasNextKin rdf:datatype="http://www.w3.org/2001/XMLSchema#string">
         Jogh Health</hasNextKin>
      </PersonalDetail>
   </hasPersonalDetail>
   <performAction>
      <Call999 rdf:ID="Call999_forFireEngine">
        <activatedBy> <AlarmFired rdf:ID="AlarmFired_6"/> </activatedBy>
      </Call999>
    </performAction>
    <hasADLHabit rdf:resource="#MakeABreakfastTea"/>
    <atLocation>
      <Location rdf:ID="Location_6">
        <withinLivingSpace> <Kitchen rdf:ID="Jemma_Kitchen"/></withinLivingSpace>
        <onCoordinates>
           <LocationCoordinate rdf:ID="oven_Coordinate">
           <hasZCoordinate rdf:datatype="http://www.w3.org/2001/XMLSchema#float">
            30.0 </hasZCoordinate>
        </onCoordinates>
       </Location>
     </atLocation>
</Inhabitant>
```

Fig. 9.4 A fragment of the OWL representation of the inhabitant instance

of the sensor. For example, the two states of a contact sensor in a microwave could be open/close. If the contact sensor is attached to a milk bottle, the literal might be used or unused. The conversion of numerical values to descriptive terms is to facilitate interpretation and comprehension for both humans and machines. Pre-processed data can then be automatically attached to semantic instances of the corresponding data source to create a data repository.

9.4.3 Semantic Content Storage and Retrieval

Once semantic data are generated, they can be archived in semantic repositories for later exchange or consumption by various applications (e.g., mining and integration). Semantic repositories are essentially knowledge bases consisting of millions of RDF triples. They are built on top of traditional database management systems by adding a semantic processing layer for semantic manipulation. Several semantic repository technologies such as Apache Jena Fuseki [13] and Neo4j [14]. are available, which could be inspiring and motivating for SSH.

Repositories may be centralised in one location or distributed in geographically dispersed sites. As all repositories share the same model, i.e. ontologies, and often use the same type of access APIs, there is little difference in the retrieval of semantic data. Nonetheless, distributed repositories are required to deal with issues pertaining to security and communication bandwidth. Within SH based assisted living, data may be exchanged and shared between institutions in different countries at a global scale. It would be desirable for each institution to have a repository and its own authorisation and authentication control for the enforcement of local data usage policies and ethical issues. On the other hand, as the volume of various data in a single SH is expected to be reasonably low, a centralised repository should be cost effective and easy for management. Therefore, it is suggested that the SSH infrastructure adopts distributed repositories at the inter-institution level and a centralised repository within an institution.

A centralised repository may be conceptually divided into two interlinked components, as shown in Fig. 9.5, based on the nature of SH data. The first component contains semantic descriptions relating to the various devices, inhabitants, individual SH and the services offered within an institution. These entities and their semantic descriptions are relatively stable for a care institution, i.e. static data. This component can functionally serve as a registry so that new SH once built within the institution, devices once added to any individual SH, inhabitants once they take residence in a SH and new services once developed can all be registered for later discovery and reuse. The second component is dedicated to the storage of dynamically generated sensory data and derived high-level ADL data, which are time-dependent, varying and extensive, i.e. dynamic data. Static data only need to be described and recorded once while dynamic data have the requirement to be recorded whenever they are generated. The separation of their storage saves storage space and also increases recording efficiency. Another advantage with this design is its ability to supports dynamic, automatic discovery of devices, device data, services and inhabitants, thus facilitating reuse of data and services. Further details of these concepts will be presented in the following section.

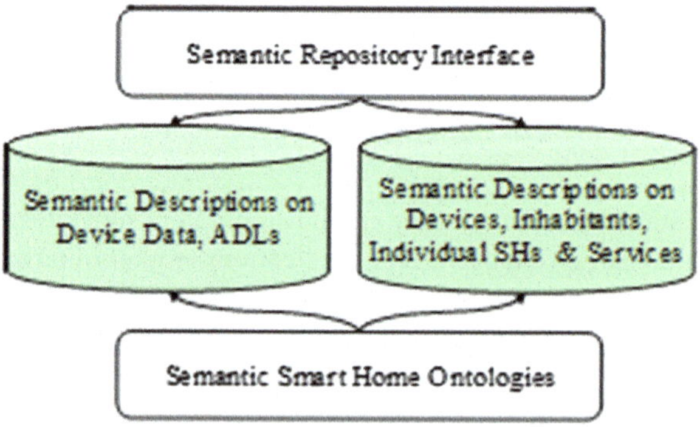

Fig. 9.5 The semantic repository within the SSH

9.5 Semantic Enabled Intelligent Assisted Agent

As semantic data are machine understandable and processable, the assistive system is able to use an intelligent agent to automatically interpret situational data for activity recognition. The Assistive Agent is responsible for the interpretation of the significance of perceived data and the provision of decision support for just-in-time ADL assistance. It performs reasoning against domain knowledge and subsequently advises corresponding actions to inhabitants. In the context of situation-aware assisted living, domain knowledge such as ADL and user profiles is formalised as Description Logic (DL) based formulae in the form of subject-predicate-object triples, e.g. the event "*FireAlarm*" leads to "*leadTo*" the action "*Call999*". They can be described in ontological relationships and represented in RDF[1] or OWL. The perception of an event and/or the detection of sensory signals are equivalent to the identification of a concrete instance of a class. For example, the activation of a contact sensor in a cup means that the cup, as an instance of *Container*, is used in an ADL. Suppose the *Container* class is the range of the *hasContainer* property, it can be inferred that the *hasContainer* property is assigned the value cup. If the *hasContainer* property is used to describe the *MakeDrink* class, it can be further inferred that a *MakeDrink* ADL has taken place. In this way the sensing of an agent amounts to the retrieval of the situational data periodically from the semantic repositories.

Central to situation-aware ADL assistance is the comprehension and reasoning capabilities of the Assistive Agent. In terms of the nature of a SH's situations the Assistive Agent can be internally designed in a two-layer framework. The Reactive Layer is used to deal with emergency situations such as an alarm fires or a pre-defined action takes place such as taking medicine at a specific time. Such situations usually involve fewer sensor data but require quick responses. The Deliberative Layer is responsible for the recognition of complex non-emergency situations that involve multiple sensory inputs. For example, sensors attached to a milk bottle, a kettle and a cup have been activated within a short time interval, how to decide the situation and further to assist the inhabitant with the completion of the ADL being performed.

An Assistive Agent comprehends perceived situational data by interpreting the data against their ontological context, i.e. ontologies. For instance, a smoke sensor in a lounge can be semantically described using two property-value pairs—[*hasConsequence, fire*] and [*hasLocation, lounge*]. Whenever the sensor is activated, an agent can interpret the occurrence against the above semantic context in the ontologies and recognize the situation "a fire breaks in the lounge". With recognized situation the future states of a SH can then be predicted and ADL assistance is subsequently provided through reasoning and inference. For example, a fire event can be semantically described with three property-value pairs—[*takeAction, toEvacuate*], [*takeAction, callFireEngine*] and [*hasEffect, homeEvacuated*]. Whenever a fire event is detected, the agent can reason against the above knowledge to advise inhabitant to evacuate the home and call fire engines. It can further deduce that the

[1]RDF and OWL are W3C standards. Detailed information can be found at W3C web site—www.w3.org.

home is empty. Reasoning at the Reactive Layer can be directly realised via built-in entailment rules in DL based ontologies.

A single sensor input can sometimes decide a specific situation, in particular for those emergency situations as discussed above. Nevertheless, most situations may involve perception inputs from multiple sources. In this case, a situation requires joint formation and interpretation of multiple perceived sensor data. For example, if sensors attached to a milk bottle, a teabag and a cup have all be activated within a short period, by linking what have happened it is reasonable to assume a situation that involves cup, sugar, milk and tea. It is straightforward for humans to figure out that this is a situation in which "*MakingTea*" ADL takes place. However, for software agents to recognize the situation as humans do, it requires an explicit representation of these situations and reasoning mechanisms. The reasoning mechanism will combine all sensor inputs to derive the corresponding situation by interpreting the aggregated perceived data against the abstract knowledge representation.

As an ADL can be viewed as a sequence of situations along the temporal dimension, we can model situations through semantic ADL modelling, i.e., to build an ADL ontology as discussed in Chap. 3. The ADL ontology consists of an ADL hierarchy in which each node, also called as a class, denotes a type of ADL. Each ADL class is described with a number of properties and sub-classes can inherit all properties from its parent class. A property is defined by specifying its domain and range. The domain refers to all classes that can be described by the property and the range refers to all classes whose instances can be assigned to the property. A property describes a class using either a literal or an instance of another class as its value, thus linking two classes. This essentially gives rise to a description-based activity/situation model, i.e. an ADL/situation is described by various properties. The underlying idea is that if a number of properties can be identified and linked, then the corresponding situation and ADL can be inferred.

The agent monitors and collects perceived sensor inputs by periodically retrieving semantic situational data from semantic repositories. These situational data have already been enriched with ontological relationships, thus ready for reasoning. The agent performs reasoning at the Deliberative Layer to derive the situation and its corresponding ADL. The process is as follows: Sensor inputs are used to identify concrete items that have been involved in ADLs. These items should have already been specified as instances of classes in SH ontologies. In terms of the scope of a property's range, the property that takes the identified item as its value can be inferred. In terms of the scope of a property's domain, the ADL(s) that can be described by the inferred properties can then be recognized. As properties can be inherited from super-classes (higher level abstract ADLs) to sub-classes (lower level concrete ADLs), the lower a class is in the ADL class tree the more properties it has. This means that the more sensor data that are available, the more accurately ADLs can be recognized. Conceptually the process amounts to the gathering of multiple sensor data at a specific time to form a situation. The situation is then used to identify the corresponding ADL and further identify these items in order to complete the ongoing ADL. This is actually a DL-based subsumption reasoning whose theoretical foundation, algorithm and implementation can be found in Chaps. 2 and 3.

9.5.1 An Example Case Study

We use the Kitchen ADL class hierarchy in 9.4.1 to delineate how our approach works. As can be seen, *KitchenADL* is the top class of kitchen ADL with two properties—*inLocation* and *HasActor*. It has two subclasses, *MakeDrink* and *MakeMeal*. Apart from inherited properties, *MakeDrink* has a property of the class Container that could be a cup, a mug or a bowl. Similarly, *MakeDrink* has two subclasses, *MakeHotDrink* and *MakeColdDrink* and each with some more properties. For example, *MakeHotDrink* ADL has two properties of the class *HotDrinkType* and *Addings* respectively. The *HotDrinkType* can assume one of tea, coffee or chocolate and the *Addings* can assume *sugar* and *milk*. Situation recognition that is denoted as corresponding ADLs is performed as follows:

We have implemented the proposed approach to situation-aware ADL assisted living in a feature-rich prototype assistive system. Figure 9.6 shows the front-end interface of the system. The system is developed with C# language as the scripting language while the front-end is developed using ASP.NET with Ajax and Silverlight support for better user experience. We use the SemWeb semantic library for C# [15] to read and write RDF, manage RDF in persistent storage, query persistent storage via simple graph matching and SPARQL, and make SPARQL queries to remote endpoints. SemWeb provides built-in general-purpose inference, but we use an implementation of the Euler proof mechanism for reasoning [16]. Euler is an inference engine supporting logic-based proofs. It is a backward-chaining reasoner enhanced with Euler path detection.

The system works as follows: A user first logs into the system and uploads the SH ontologies from the BASE ONTOLOGY panel. By registration and login page, the user establishes his/her identity. As such the user's ADL preferences can be browsed in the USER PREFERENCES panel. Once SH ontologies are loaded, the system can display sensors that are semantically described. At this stage the system can operate in two modes—simulated and real-time ADL monitoring. In the simulated

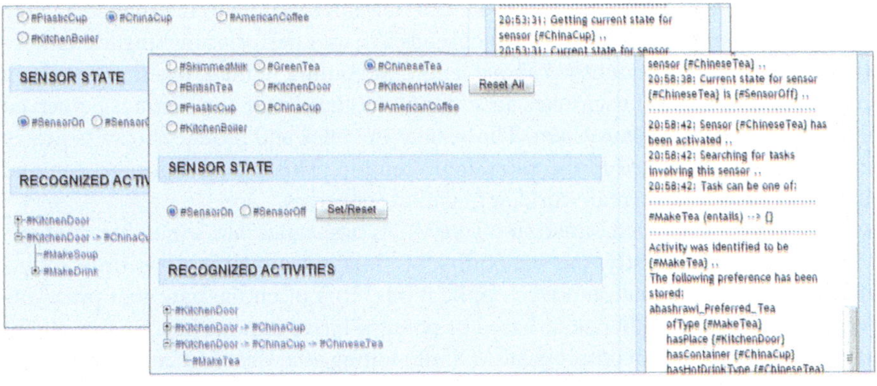

Fig. 9.6 The incremental situation formation and ADL recognition process

scenario, the system does not need to be connected to sensors. Sensor activation is simulated by the selection of a sensor, e.g. KitchenDoor, in the SENSOR SOCIETY panel and the set-on of the sensor in the SENSOR STATE panel, see Fig. 9.6. This is equivalent to the activation and perception of real sensors. Once a sensor activation is observed, either simulated or triggered in real time, it will be used to form a situation to reason against the semantic ADL descriptions. The LEARNING OUTPUT panel displays the inference process of the assistive agent as sensors are activated and events perceived. The RECOGNISED ACTIVITIES panel displays the recognized ADL and its location in an ADL tree structure. Both are shown in Fig. 9.6.

Figure 9.6 illustrates the dynamic situation formation and incremental ADL recognition process. When a *KitchenDoor* sensor is activated, only high-level ADL such as *MakeMeal* and *MakeDrink* can be inferred—see c. When *ChinaCup* and *ChineseTea* sensors are activated sequentially, situations with more contextual details can be dynamically formed. By reasoning these situations an assistive agent can recognise the ongoing ADL progressively in increasing details, e.g., *MakeDrink* initially and then *MakeTea* as depicted in Fig. 9.6.

Suppose that a user has a pre-defined, semantically described preferred ADL *MakeTea*. By comparing the user's *MakeTea* profile with the perceived situation, an assistive agent can infer what shall be done next in order to complete the ongoing ADL, thus providing situation-aware ADL assistance for users. For example, if "*abashrawi-preferred-tea*" ADL contains *sugar*, the agent may remind the user to add *sugar* if it does not detect the activation of the sugar container for a pre-defined period of time.

On the other hand, if a user activity has been recognized repeatedly over a relatively long period of time, and there is no corresponding matching ADL profile, the activity can be recorded as a user's preferred ADL profile. This is the learning process. We shall not discuss it here in details due to space limits.

In addition, to evaluate the approach and system in the simulated scenario, we have designed an experiment in our smart home environment—see Fig. 9.7, for evaluation of the proposed approach and the implemented system in a real-world use case. We attach contact sensors to teabag, sugar, kettle, milk and cup containers—see Fig. 9.8. Then we connect the prototype ADL assistive system to the sensors via the Tynetec wireless receiver. The experiment runs as a user performs making tea activity following the scenario discussed above, i.e., first coming to the kitchen, then taking a cup, etc. Each time the user takes an action/item, the sensor activation is perceived and passed to the assistive system. The system operates and produces results nearly the same as we discussed in the simulated scenario. This proves the approach and system are applicable in real-world application scenarios.

Semantically enhanced situation-aware ADL assistance has a number of compelling advantages: Firstly, the scalability of situation modelling has been a bottleneck to effective situation-aware applications. It is often the case that proof-of-concept experiments, either state-based or process-based approaches, work well but fail to scale up. The use of ontological ADL modelling as a way of situation modelling overcomes this problem. Ontology engineering offers extensive technological support, including tools, APIs, storage and reasoners. Ontologies of thousands of classes have been developed in other domains, e.g. 7,000 concepts in the gene ontology, and

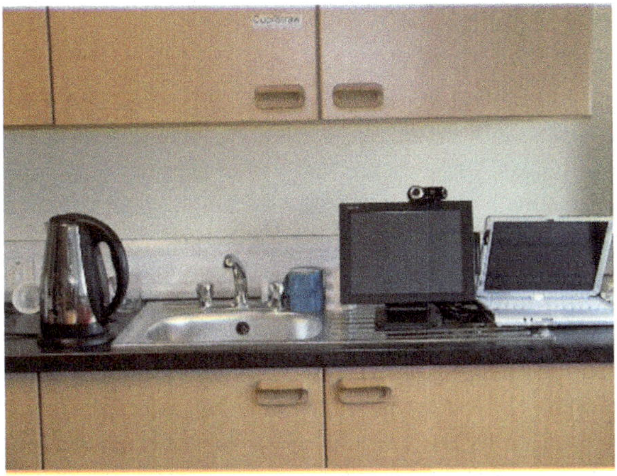

Fig. 9.7 The smart home and connected assistive system

Fig. 9.8 Sensor network design in the experiment

semantic data repository of 25 million triples has been practiced in TripleStore [17]. For smart homes, ADL classes and associated instances are simply not present in such a scale. Secondly, semantic ADL models contain explicit rich semantics and built-in logical entailment rules. This allows not only humans but also assistive software agents to interpret, comprehend and reason against semantic situational data. As such, situation monitoring, and ADL recognition can be realised at higher levels of automation. Thirdly, description-based reasoning provides a mechanism to dynamically construct situations by interpreting limited or incomplete sensor data that ultimately leads to the incremental recognition of the corresponding ADL. This capability is particularly important because assistive systems are supposed to provide reminding or suggestive assistances with limited sensory data.

9.6 Summary

The SSH concept, its enabled assisted living paradigm and underpinning technologies await further investigation, development and evaluation through real-world use case studies. Nevertheless, our work has laid a solid architectural and methodological foundation. Initial results have demonstrated the potential and value of the approach and further clarify future research directions. We believe that SSH are the next generation of technological infrastructures for assisted living that facilitates the innovative exploitation of research results from AI, Web technologies and information processing. In this chapter a semantic-enabled agent-based novel approach is discussed for enhanced situation-aware assisted living assisted living. We have discussed the concept of situation awareness and introduced an integrated system architecture for semantically enhanced situation awareness and intelligent just-in-time ADL assistance provision. We have analysed the nature and characteristics of SH-based assisted living assisted living. Based on the analysis we describe semantic data management including SH ontologies, semantic data creation and storage. We have presented the use of assistive agents for situation comprehension and ADL recognition with special emphases on the agent's internal structure and its interpretation and reasoning mechanisms. A simple yet convincing example scenario from a real-world ADL assistance context has been used to illustrate our approach.

We have implemented a prototype assistive system for the proposed approach using the latest semantic technologies and toolkits. We have carried out both simulated and real-world use case study. While the full evaluation of the proposed approach and system awaits further extensive implementation and user feedback, initial research results have been promising. Our future work aims to address temporal issues such as parallel/concurrent ADL recognition. We shall extend the existing assistive system with capabilities of taking actions, e.g., playing audio/video or switch on/off devices/appliances through actuators.

References

1. Boger J, Poupart P, Hoey J, Boutilier C, Fernie G, Mihailidis A (2005) A decision-theoretic approach to task assistance for persons with dementia. In: IJCAI international joint conference on artificial intelligence
2. Bouchard B, Giroux S, Bouzouane A (2006) A smart home agent for plan recognition of cognitively-impaired patients. J Comput 1(5):53–62
3. Jennings NR, Wooldridge MJ (1998) Agent technology: foundations, applications and markets
4. Tang Z, Guo J, Miao S, Acharya S, Feng JH (2016) Ambient intelligence based context-aware assistive system to improve independence for people with autism spectrum disorder. In: Proceedings of the annual Hawaii international conference on system sciences, pp 3339–3348
5. Meditskos G, Kompatsiaris I (2017) iKnow: ontology-driven situational awareness for the recognition of activities of daily living. Pervasive Mob Comput 40:17–41
6. Zolfaghari S, Zall R, Keyvanpour MR (2016) SOnAr: Smart Ontology Activity recognition framework to fulfill Semantic Web in smart homes Samaneh. In: Proceedings of the annual hawaii international conference on system sciences, pp 3339–3348

7. Sycara K, Paolucci M, Lewis M (2003) Information discovery and fusion: semantics on the battlefield. In: Proceedings of the 6th international conference on information fusion, FUSION 2003
8. Laskey KB, Haberlin R, Costa P, Carvalho RN (2011) PR-OWL 2 case study: a maritime domain probabilistic ontology. CEUR Workshop Proceedings, vol 808, pp 76–83
9. Endsley MR (1995) Toward a theory of situation awareness in dynamic systems. Hum Factors J Hum Factors Ergon Soc 37(1):32–64
10. Noor MHM, Salcic Z, Wang KIK (2018) Ontology-based sensor fusion activity recognition. J Ambient Intell Humaniz Comput 1–15
11. W3C: The time ontology in OWL. https://www.w3.org/TR/owl-time/
12. Stanford University, University, S.: Protégé
13. Apache: Apache Jena. https://jena.apache.org/
14. Neo4j: Neo4j graph platform – the leader in graph databases. https://neo4j.com/
15. Semantic Web: semantic Web/RDF liberary for C#.NET. http://semanticweb.org/wiki/SemWeb-DotNet.html
16. W3C: Euler proof mechanism. http://www.agfa.com/w3c/euler/
17. Harris S, Gibbins N (2003) 3store: efficient bulk rdf storage. In: Proceedings of the 1st international workshop on practical and scalable semantic systems (PSSS'03)

Chapter 10
Human Centred Cyber Physical Systems

10.1 Introduction

Previous chapters introduce core underlying technologies for human activity recognition, covering approaches, models, methods, algorithms and mechanisms for the main functionalities of the lifecycle of activity recognition. They also cover different scenarios, i.e. single-user sequential activities, single-user interleaved or concurrent activities and multi-user collaborative activities, focusing particularly on knowledge-driven approaches. In undertaking the research under various application scenarios, four human-centred smart cyber-physical systems for assisted living have been developed to test and evaluate the research outputs. This chapter will describe the four cyber physical system (CPS) prototypes and their key components, features, applications and the technologies in detail. The first two systems follow a standalone system architecture while the latter two on a distributed system architecture.

The first standalone system, dubbed as SMART [1, 2], was developed with a direct sensor interface to the SH environment and feature rich web-based human-machine interface using dotNet programming language. The SMART system consists of six functionality classes, namely speech core, reasoning core, preferences core, communication core, simulation recording core and database management core. The speech core class is used to output pre-recorded audio messages to the user when the assistance is triggered, which also support pre-recorded personalised messages. The reasoning core and preferences core classes are the core components of this system. The reasoning core class is used to infer the user's activity from their preferences. The user preferences are administered via basic or advance learning methods presented by the system. The sensor's activation data are stored in the database management core and retrieved via the communication core. The data from the sensor activations, inferred activities from reasoning can be recorded using simulation recording core class. This data can then be exported to the user's local disk or stored in the repository database as a history log.

The second standalone system, namely agent-mediated AR system, introduce modular components that work in parallel. This prototype investigated methods to

L. Chen and C. D. Nugent, *Human Activity Recognition and Behaviour Analysis*, https://doi.org/10.1007/978-3-030-19408-6_10

recognise composite activities such as interleaved and concurrent activities. The approach combines the recognition of single and composite activities into a unified framework. To support composite activity modelling, it combines ontological and temporal knowledge modelling formalisms. In addition, it exploits ontological reasoning for simple activity recognition and qualitative temporal inference to support composite activity recognition. The approach is organised as a multi-agent system to enable multiple activities to be simultaneously monitored and tracked.

In the third implementation, the SOA approach was introduced with open source components. This approach essentially follows a client-server pattern, to resolve some of the technical challenges mentioned above in building an assisted system within SH environments. The key benefit of this approach is that it allows one or more clients to communicate with the SMART system simultaneously irrespective to their operating platforms. SOA can take advantage of cloud computing to increase the scalability and resources to perform complex reasoning or computation tasks with less time [3].

The core system of the third system was written in a popular Java programming language. The main reasons were to move away from a standalone environment, limited community support and proprietary components. This approach allows multiple users from multiple devices to communicate simultaneously with platform independence. The system further addresses the monolithic code structure of the source code by logically separating it into three web services. The Enterprise Service Bus (ESB) is used to bind these services together, enabling better maintainability, reuse and debugging. The system still has a web-based interface that uses JavaScript, Asynchronous JavaScript and XML (AJAX) features to request and load data from the ESB. In addition, the Simple Object Access Protocol (SOAP) and Hypertext Transfer Protocol (HTTP) has been used for exchanging data between different devices. One of the disadvantages of using this system is that it has multiple web services with an ESB which requires it to be hosted on the network. This can create unnecessary overheads and delays to the system.

The last prototype extends SOA SMART implementation by proposing multi-layer system architecture and Representational State Transfer (REST) based web service. The REST-based web service allows communication between clients or sensing devices with lower overhead and energy consumption. The multi-layered web service enables components to be decoupled and deployed as single instances on the server. The multilayer approach caters for increasing features to be added to the system by organising and decoupling components for a better reusability and maintainability. In comparison to ESB approach in third prototype, multiple web services are required to be hosted on the network which can be subjected to additional overheads and data transmission delays. The sensor events or results in previous SOA implementation were stored in traditional RDBMS compared to graph-based database in multi-layered system. The graph-based database enables semantic metadata to be extend the new connections and attributes infinitely instead of modifying existing table schema and redefining entity relationships between tables in RDBMS. The sensor event processing in previous SOA system adapted continuous querying approach with RDBMS and then perform semantic reasoning tasks. On the other hand, the multi-layered system inspects sensor events as received and segmented at run time. The data is stored using graph-based database to preserve the semantical metadata of the sensors and AR results. A multi-threading process enable progressive reasoning to be performed as activities unfold. The incremental reasoning feature from Pellet is leveraged to reason with only the affected changes. This process reduces bespoke querying effort to obtain sensor data from traditional database and then perform semantic reasoning with whole ontology file. Finally, Android application and web interface has been developed to utilise the lightweight web service to view sensor events, AR results, manage user preferences, and other features. Compared to third prototype, the key limitation of the fourth prototype is the run-time memory and number of cores required to conduct all the data collection, reasoning and storage tasks on a single machine.

10.2 SMART: A Standalone System for Sequential Activity Recognition

SHs have emerged as a commonly agreed, technically viable solution for ambient assisted living. While early work concentrated on physical SH prototyping, supportive technology development and the experimentation with activity assistance, trends are now moving towards context modelling, activity recognition and cognitive assistance. Though progress has been made in individual areas, an optimum solution offering a scalable, flexible, easy-to-deploy solution has not been produced which can assist a SH inhabitant to perform the "right" activity at the "right" time in the "right" place. Specifically, a SH is expected to be able to (i) monitor the behaviour of the inhabitant, (ii) learn his/her preferences and (iii) provide assistance on what, where and when an activity is performed.

In our research, we have developed a function-rich cyber-physical system for assisted living that has the potentials of assisting people to perform the right activities at the right time and in the right place. The system is able to monitor the behaviour of an inhabitant, learn his/her preferences and provide context-aware assistance. It is based on formal ontological activity models that make use of markup languages for semantic and knowledge modelling and utilises their expressive representation power for reasoning. The resulting ontologies are essentially shared knowledge models that facilitate interoperability and integration in terms of the shared structure and terminology. They enhance automatic processing and the level of automation. The proposed reasoning algorithm can not only deal with every sub-activity as an isolated concept and carry out reasoning but also use knowledge acquired from previous sub-activities to obtain the most specific activity being performed by the user. By grouping these concepts together, the system can achieve a more accurate outcome with less processing time. Moreover, the system can learn from a user's behaviour and adapt itself according to the user's activity profile; such as enabling personalized assistance based on the way the activity is performed before. We describe the system in detail below, so researchers interested can use the system or follow the rationale for their own research.

10.2.1 The SMART System Architecture

Figure 10.1 shows the conceptual architecture of the SMART system. The bottom layer data model contains the ontological activity model which are shared knowledge across the commonly shared terms in the problem community. This shared knowledge is separated from the preferences specific to individual users. As can be seen in the architecture, users interact with smart devices present in the SH and this information needs to be monitored by the system and processed for activity recognition. Based on the ADL ontology model, activity(s) matching the input from the user are specified. Once this is done, assistance is provided by comparing sensors activated so far in this activity against the sensors found from this user's preferred way of conducting this activity as found in the ADL ontology. The list of sensors not used yet is carried to the user in the form of assistance. Meanwhile, the system should update preferences when necessary to adapt with users' changing lifestyles. Table 10.1 gives more detailed information about this concept.

The main components and their relationships among environmental objects, sensors, ADLs and users are shown in Fig. 10.2. Central to this knowledge representation is the ontological modelling and representation of SH domain knowledge in the left column. The Context Ontologies are used to semantically describe contextual entities

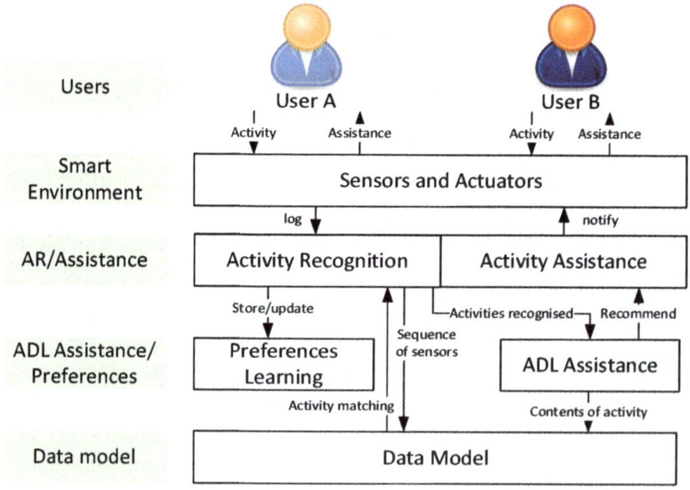

Fig. 10.1 The SMART system architecture

Table 10.1 Description of different layers in the conceptual model

Layer	Information
Users	Users will interact with the sensors and expect to receive assistance on what is the activity being performed and what to do next
Smart environment	The smart environment consists set of sensors and actuators to enable monitoring of the user activities and notify for any assistance
Activity recognition/assistance	Receives a list of sensors activated and uses the ADL ontologies to find activities matching the list. Using ADL Assistance/Preferences, it conducts assistance needed to end users
ADL assistance/ preferences	Receives activity(s) identified and uses the ADL ontologies to find sensors associated with the activity(s) identified. The set of sensors not activated yet is carried to the Activity Recognition/Assistance. New knowledge obtained about the user's preferred way of conducting this activity is updated in the ADL Ontology Model
Data model	Models activities and preferences in a hierarchical structure. Explicitly specifies key concepts and the relationships among them for a problem domain. Also splits common knowledge from specific knowledge

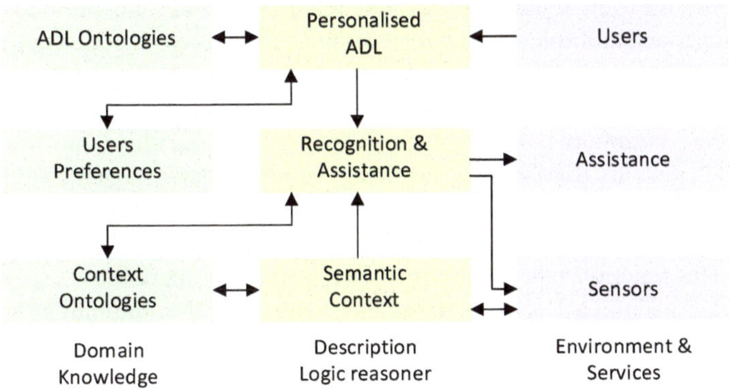

Fig. 10.2 The relationships between various entities within a SH

and sensor observations. The generated semantic contexts, i.e. situations are used by the Recognition and Assistance component for reasoning, activity recognition and assistance. The ADL ontologies are used, on one hand, to create ADL instances for an inhabitant in terms of their ADL preferences, and on the other hand, to serve as a generic ADL model for activity recognition and coarse-grained activity assistance.

The components in the right column denote the physical environment, sensors, devices and assistive services in a SH. The sensors monitor an inhabitant's ADL and use their observations, together with context ontologies, to generate semantic contexts. Assistive Services receive instructions from the Recognition and Assistance component and further acts on the environment and/or the inhabitant through various actuators.

Activity recognition and assistance is performed through a description logic-based reasoner as shown in the middle column. The reasoner takes as inputs the semantic descriptions of a situation and performs reasoning against the ADL ontologies to provide incremental progressive activity recognition and coarse-grained ADL assistance. To support fine-grained assistance, concrete sensor observations will be bound with context models to create an activity's description. By reasoning the descriptions against an inhabitant's personal ADL profile, specific personalized assistance can be suggested and passed onto Assistive Services in a SH.

10.2.2 The SMART System Implementation and Operation

SMART is a feature-rich context-aware assistive system as shown in Fig. 10.3. It is developed using C#, ASP.NET, Ajax and Silverlight support for audio and graphical user experience. The creation, management and query of the semantic data was handled using the SemWeb semantic technologies for C# [4] and SPARQL [5]. Semantic reasoning was implemented using the Euler inference engine [6].

To support the rich functionalities and features, the system provides a set of configuration tools, multiple graphical views and quick-access function buttons (refer to Fig. 10.3). These tools are used to import activity and context ontologies, specify reasoning and learning parameters, select the modality of audio reminder, configure hardware, e.g., communication ports, and define event priorities and user activity profiles. The views are used to show the deployed sensor network in the environment, display the sequence of activated objects, output a temporal trace of events during the system operation and present the recognised activity within a tree-like activity hierarchy. The function buttons allow a user to initiate operations, e.g., to record sensor activations in XML files or to put the system into the simulation mode using previously recorded sensor activations in a XML file.

The system was deployed in a computer in a smart home laboratory [7]. The spec of the computer used was as follows: Dell Optiplex755, Intel Duo CPU E6750 2.66GHZ, 3.25 GB RAM, Windows XP SP3. The SH environment used for experimentation contains various objects for performing ADLs. A number of sensors, including contact sensors, motion sensors, tilt sensors and pressure sensors, all from the Tynetec ZP0532 series, are available. A Tynetec receiver (www.tynetec.co.uk) is used to collect sensor activation signals through wireless communication which is subsequently processed by the computer system.

When the system is in operation, it obtains real-time sensor activations from a designated communication port that is connected to an external Tynetec receiver. Each time a sensor is activated, it will aggregate the information with previously collected activated sensors to generate an activity description. The description is then fed to the reasoning engine to infer the potential activity against activity models and profiles.

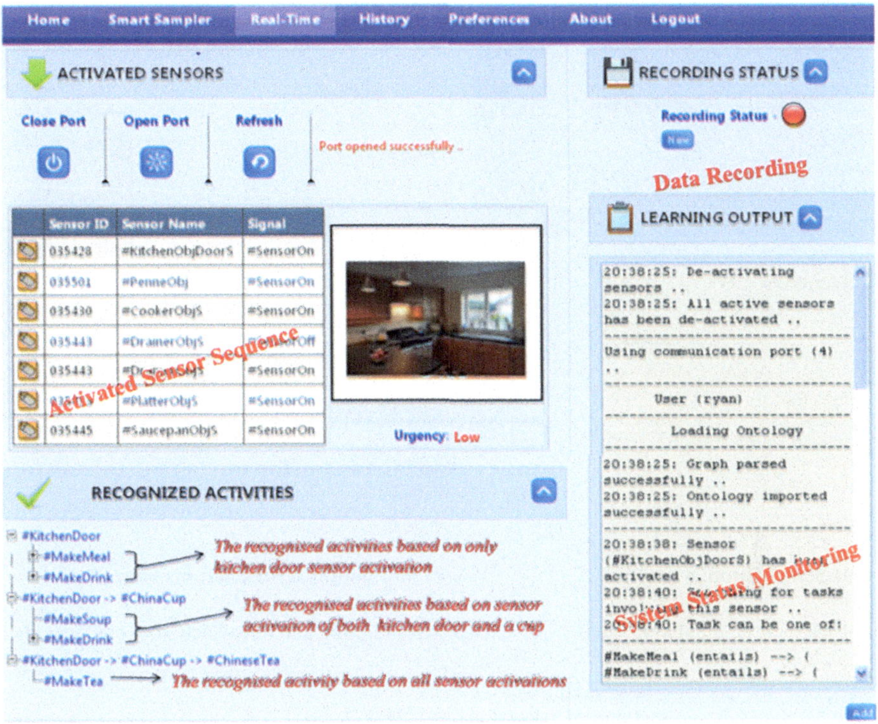

Fig. 10.3 The system interface in real time mode

As an actor interacts with the objects in sequence in real time, sensor activations are continuously fed into the system. As such, recognition operations are repeatedly performed to realise continuous progressive activity recognition. As can be seen in Fig. 10.3, the system can dynamically display the activated sensor sequence, the incrementally recognised activities, and system status and data. Figure 10.4 enables the user to browser different sensors located in the house (a), update their status activate and store/open the simulation activities from the local disk.

10.2.3 SMART Limitations and Opportunities

The SMART system implementation has four disadvantages. One disadvantage is the limitation imposed on technologies that can be efficiently used to produce the system. More specifically, as the standalone system is developed using dotNET technologies it is limited to using components either developed for dotNET or implementation technologies that have a software compatibility layer developed for them, this is generally inefficient [8]. Many semantic ontology modelling and reasoning tools are

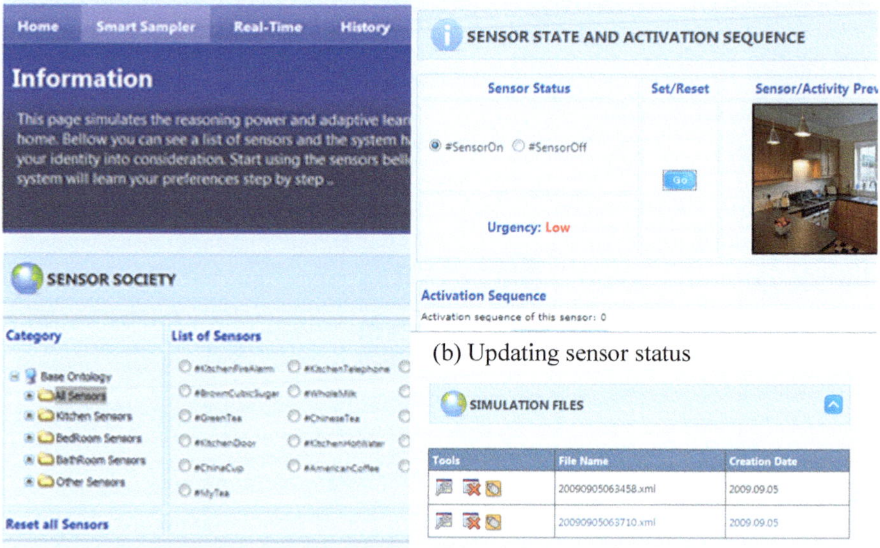

(a) Sensor simulation

(b) Updating sensor status

(c) Simulation scripts/actions stored

Fig. 10.4 Sensor simulation interface to activate sensors, store and reload actions from local disk

not developed in dotNET technologies [9] and so will not be able to operate with (e.g. PERL has no such method to directly interoperate with Objective-C running in a different address space [10, 11]) or efficiently integrate with the current system.

In order to increase the semantic modelling capability of this system more advanced semantic ontology tools such as Pellet or Fact++ would need to be incorporated in an efficient manner [10]. The use of web services would allow expansion of this system using components containing these advanced tools using a native technology hosted web service.

The second disadvantage is that for every instance of a SMART environment functionality is duplicated and hosted locally. In a monolithic code base, there is a need to provide a complete system for each assisted living environment. A better solution would be to move some functionality to a shared resource which is accessible via web services and so would allow Internet Protocol enabled sensing and interaction equipment to communicate with external web services. Use of web services in this fashion would remove the need to have a locally hosted assisted living server computer.

Web services have an additional advantage related to their shared nature, in the monolithic deployment scenario if a software update is developed for the assisted living service software needs to be applied to each instance of the software (each computer hosting this system in a residence). A mass update may produce issues depending on deployment mechanism or simply if there is a power outage during the update of a critical file. In a web service orientated system, the software update would simply need to be applied to the relevant web service node [12].

The third limitation is the monolithic code structure has a potentially constrained future development lifecycle. A large single code base requires coordination between developers who are expanding system functionality in different components as the final compiled product has a single address space any changes to a component has the ability to affect other components potentially leading to corruption in the operation of other components. It would be better if developers of components could work on a single component in its own address space and offer an accessible interface to its functionality, this is the facilitated by use of web services. Web services also introduce an additional aspect of software reusability as any existing published service functionality (e.g. authentication) can be used without writing the business logic to do so.

The fourth disadvantage is that this SMART system requires a licencing fee for its hosted OS. The standalone system is based on the Microsoft dotNET framework which is only fully supported on Microsoft Server platforms (IIS) which carries a licencing fee in order to legally use them [13]. The standalone system implementation does not require connectivity with hosted web services. Being a locally hosted monolithic system there is no requirement to be able to communicate with the web service components and so a monitored environment would need to ensure that connectivity with the relevant web service nodes are guaranteed.

This initial SMART system can be enhanced by adopting agent or SOA approach to remove all dotNET technology to avoid licensing costs. The current system does have one notable advantage over a web service-based solution. In a scenario where many residential units are connected to a locally hosted web service that does not require internet access, this should not introduce much of a problem. However, if the local assisted living equipment connects to an internet hosted web service components then the functionality of the assisted living environment is not guaranteed, in such cases redundant networking would help to ensure connectivity and system functionality [14] (e.g. use of a combined cellular and DSL gateway).

10.3 An Agent-Based System for Composite Activity Recognition

Composite activity recognition can be divided into three interdependent sub-tasks, namely, action recognition, simple activity recognition, and composite activity recognition. Action recognition processes the sensor data available in sensor data segments using ontological reasoning to derive primitive actions. The action recognition task is performed as part of simple activity recognition. For simple activity recognition, we adopt and modify the ontological activity recognition approach described in Chap. 3. Ontological activity recognition uses a logic-based ontology language, e.g. OWL, to structure and describe activities during activity modelling. It encodes activity models as activity ontologies, and then uses semantic reasoning (e.g. subsumption and equivalence reasoning or instance retrieval) to process sensor data against the ontological activity models during activity recognition. To support composite activity recognition, we modify the ontological approach by including a step to generate activity descriptions. We define an activity description as a collection of primitive actions that together, partially or fully, describe a simple activity. An activity description can be created by grouping the primitive actions into one or more activity descriptions corresponding to the simple activities that are defined in the activity models. As more sensor data is obtained new activity descriptions are created or the existing ones are updated. The modified ontological approach then compares each activity description with activity models using semantic reasoning and reports the activity model that is closest to the activity description as the ongoing simple activity. The activity model returned by instance retrieval is considered the closest model. In the absence of a model returned by instance retrieval, then the model returned by equivalence reasoning is taken as the closest. Otherwise, the model returned by subsumption is the closest. Thereafter, the results of simple activity recognition are aggregated using the mechanism described in the next section. By separating activity recognition into interdependent tasks, it is possible to use different techniques for each task. In this way instance retrieval or subsumption and equivalence reasoning is used for action and simple activity recognition. For composite activity recognition, rule-based inference techniques are exploited.

10.3.1 The Conceptual Architecture

The approach described above can be depicted in a modular architecture as shown in Fig. 10.5. It consists of a number of core components that interact with each other to provide intended functions. Core to the architecture is three knowledge bases (KB), namely, static activity model KB (*StatSKB*), dynamic model of composite activities KB (*DynaCAKB*), and context-driven rule-base (*ContextRB*), which are utilized by the different components during operation. *StatSKB* provides the static model of activities and includes definitions of activities of daily living as well as predefined composite activities. *DynaCAKB* encodes the dynamic model of activities. *ContextRB* encodes the rules for inferring qualitative temporal relations between activities and therefore deriving the ongoing composite activities. At runtime *DynaCAKB* and *ContextRB* are used to derive the temporal dependencies that exist among ongoing activities. The presence of temporal dependencies among activities implies the existence of composite activities. The knowledge bases are encoded as an activity of daily living (ADL) ontology. The data monitoring and segmentation component monitors and collects contextual and sensory data whenever a user interacts with objects in performing daily activities and then segments the sensor data stream.

The integrated activity inference component performs three tasks, namely iterative action inference, activity inference and activity analysis and refinement. These tasks are performed in complex activity recognition unit (CARU) and simple activity

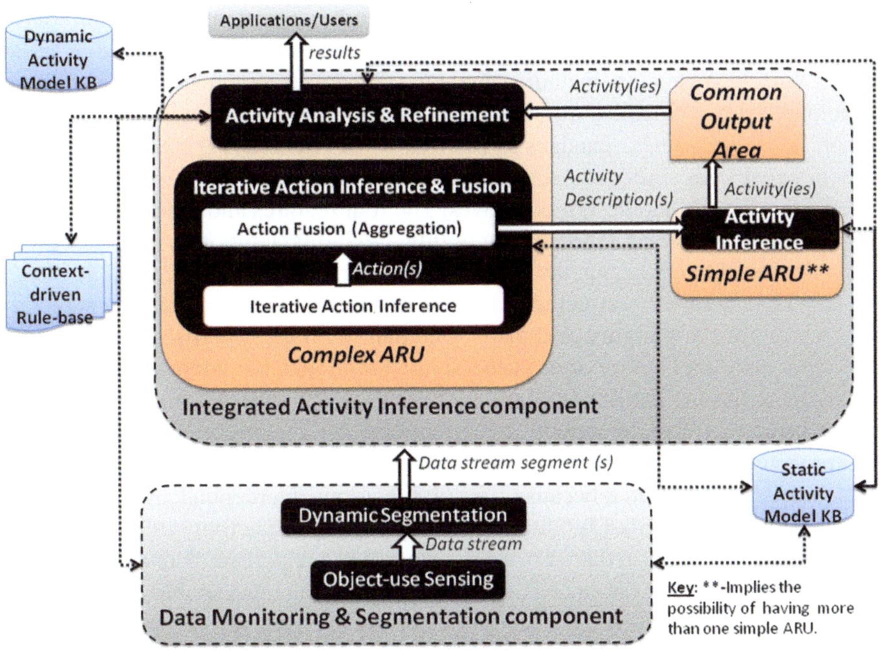

Fig. 10.5 The modular architecture of the proposed approach

recognition unit (SARU). CARU performs its task using iterative action inference and fusion, and activity analysis and refinement components based on the *StatSKB*, *DynaCAKB*, and *ContextRB*. The iterative action inference and fusion component uses the sensory information contained in the data segment and the *StatSKB* to derive primitive actions and activity descriptions. The activity analysis and refinement component is used to discover complex dependencies among ongoing activities. It uses various elements of context information, e.g. task-related context and temporal context encoded in *ContextRB*. The analysis and refinement component outputs simple activities or composite activities together with feedback that is used in the segmentation component to modify the parameters used in the segmentation mechanism. On the other hand, SARU performs the necessary activity inference autonomously and communicates its status. It uses its activity inference component to derive the activity that corresponds to a given activity description. Activity inference uses *StatSKB* as well as recognition algorithms.

The conceptual modular architecture has been realised in a multi-agent system. An agent refers to a software system that is situated in a dynamic, complex environment, and is capable of sensing the changes in the environment and interacting with other entities in order to take actions that achieve its design objectives. An agent should exhibit four basic properties, namely, autonomy, social ability, reactivity (responsiveness) and pro-activity [15]. An agent that is autonomous is able to act with no direct intervention from humans or other agents and has control over both its internal state and actions. Social ability refers to the ability of an agent to interact with other agents, including humans. Generally, an agent is situated in an environment, and the ability to perceive the environment and respond in a timely fashion to environmental changes is referred to as reactivity. Pro-activeness indicates an agent's capability to exhibit goal-directed behaviour by taking initiative to achieve its set goals and design objectives. The four features and other features (e.g. adaptability, intelligence, rationality, mobility, flexibility, temporal continuity, etc.), can be exploited to design agents for use in applications in the AAL domain.

Agents can be used to structure solutions in application areas that are characterized by complexity, ubiquity, and distributed data, control, expertise, and resources [16]. One such application area is the smart home and by extension the task of activity recognition. The smart home is characterized by various components, e.g. sensors, actuators, people, activities, and interactions, making it complex. In addition, the smart home is by definition ubiquitous. Moreover, activity recognition can be characterised as distributed because it involves various interdependent tasks such as environment and behaviour monitoring, segmentation, and activity inference. Also, each task requires its own data, resources, control and expertise. A multi-agent system (MAS) refers to a system consisting of a group of agents capable of interacting with each other to achieve their design objectives [17]. Therefore, the multi-agent

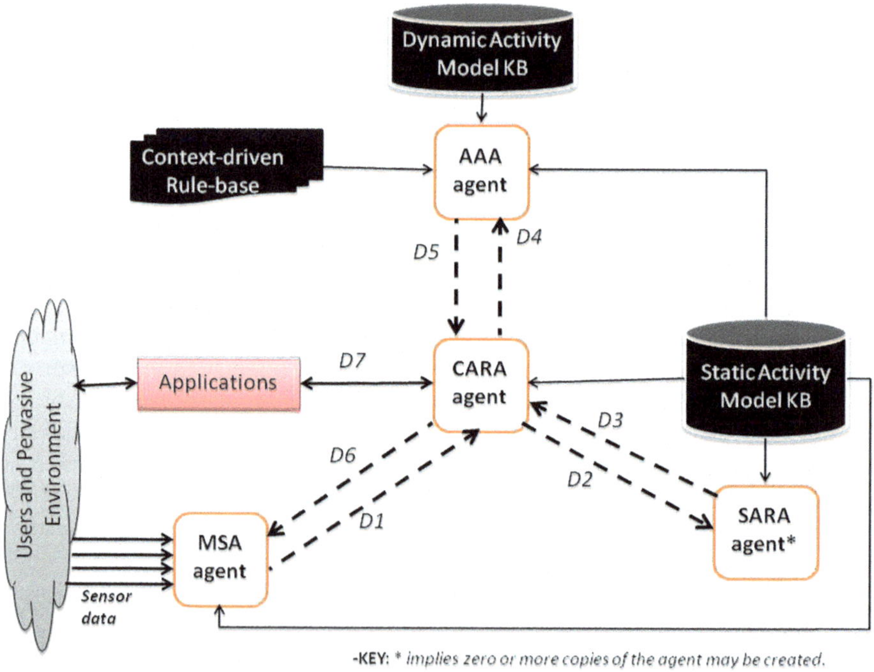

Fig. 10.6 The multi-agent architecture for unified activity recognition

approach can be used to model, structure, and implement a complex software system as a collection of interacting, autonomous agents [18]. From the foregoing, we adopt the multi-agent approach to specify a system for unified activity recognition.

To realise the modular architecture shown in Fig. 10.6, four agent roles are identified, namely, segment the sensor data stream, generate activity descriptions and convey recognition results, infer simple activities, and manage inference rule execution and infer composite activities. The resulting multi-agent system consists of four types of agents that play the roles stated and the agents include monitor and segment agent (MSA), composite activity recogniser agent (CARA), simple activity recogniser agent (SARA), and activity analysis agent (AAA). The messages exchanged between four agents convey the following: D1-data segments; D2-activity descriptions; D3-simple activity labels; D4-activity data; D5-simple and composite activity labels; D6-recognition status; D7-identified activity labels. Using the agents, the CARU component as described in Fig. 10.6 is implemented using the two agents, i.e., CARA, and AAA; SARU is implemented by the SARA agent; and the data monitoring and segmentation component implemented by the MSA agent.

We have chosen agent as an implementation artefact because agents provide the different components with autonomy needed to perform their respective tasks. In addition, each component can continuously and proactively review and react to changes in its goals. There is also massive parallelism involved in executing the various tasks involved and the MAS can implement the tasks as parallel agent behaviours or tasks. The resulting multi-agent architecture is shown in Fig. 10.6. Each of the agents in the architecture is described below.

Monitor and segment agent (MSA). The monitor and segment agent plays the role 'segment the sensor data stream'. Essentially, MSA receives streaming sensor data from the environment and uses time windows to segment the stream in real-time. It then sends the resulting segments to the CARA agent for further processing.

Composite activity recogniser agent (CARA). The composite activity recogniser agent plays the role 'generate activity descriptions and convey recognition results'. The CARA agent obtains segments from MSA agent and processes them to determine the actions entailed. Consequently, it uses the actions to generate activity descriptions that approximate the activities that are likely to be occurring. The CARA agent then spawns the SARA agents, and provides each with the relevant activity description. It will keep updating the activity descriptions and communicating the descriptions to SARA agents. In addition, it receives feedback related to activity labels from SARA agents, and conveys activity data to AAA agent. Finally, it obtains the results from AAA agent and provides results, i.e., identified simple and composite activity labels, to applications. Moreover, it sends information about the recognition status to MSA agent to facilitate dynamic segmentation.

Simple activity recogniser agent (SARA). The simple activity recogniser agent plays the role 'infer simple activities'. The SARA agent receives activity descriptions and their revisions from CARA agent and performs ontological inference to determine associated activity labels. It then conveys its recognition status—a specific or generic activity label—to CARA agent. In addition, the SARA agent continuously reviews its status and can terminate if a predefined upper temporal duration threshold is exceeded. At runtime, zero or multiple SARA agents can be created and executed, with each agent corresponding to exactly one activity description. Whenever activities are performed in parallel during a particular time interval, multiple activity descriptions will be derived, and corresponding SARA agents will be executed thus allowing the entailed activities to be recognised. The results from these multiple SARA agents are used as input to composite activity recognition.

Activity analysis agent (AAA). The activity analysis agent plays the role 'manage inference rule execution and infer composite activities'. It receives activity data from the CARA agent, and executes inference rules to determine the presence of inter-activity dependencies, e.g. sequence or concurrency. It only signals the presence of composite activities if it can determine that inter-activity dependencies exist. Finally, the AAA agent conveys the results—simple or composite activity labels—to the CARA agent.

10.3.2 Multi-agent System Implementation

The multiagent system has been implemented using Java Agents Development Framework (JADE) [19]. JADE is a Foundation of Intelligent Physical Agents (FIPA) compliant agent development environment. The basic standards for FIPA include agent communication, agent management, and agent/software integration. The standard for agent management is aimed at allowing agents to register, deregister, be searched, and be modified. The standard for agent communication is concerned with the message transport protocol, message content, and communication language.

Four types of agents have been developed to play the roles described in the previous section. Each agent advertises its capabilities by registering with JADE's Directory Facilitator so that other agents can search for it. The agents communicate with each other by exchanging messages represented as serialized objects. Each agent decides on the type of agent that should receive a particular message. In addition, the multi-agent system uses a communicative act theory to manage conversations between agents.

The ADL Ontology is designed using OWL 2 in Protégé ontology editor. This includes an implementation of the entailment rules as Semantic Web Rule Language (SWRL) [20] rules as part of the ADL Ontology. The prototype system uses Java-based application programming interfaces (APIs) to interact with the Pellet [21] OWL reasoner for ontological reasoning. To facilitate the execution of the inference rules the ADL Ontology and the SWRL rules are translated to Java Expert System Shell (JESS) [22] fact and rule bases. We used the OWL2Jess and SWRL2Jess translators based on the guidelines provided by Mei and Bontas [23]. In the prototype, the JESS fact and rules bases are accessed and processed by a JESS rule engine. The rule engine is accessed and manipulated by the AAA agent that is responsible for aggregating the results of simple activity recognition.

10.3.3 Multi-agent System Interface

The interface for the multi-agent assisted living system described above and in Chap. 7 can be viewed in Fig. 10.7. Figure 10.7a shows five sensor observations for the *MakePasta* activity. At the same time Fig. 10.7b shows the agent instances as obtained from JADE's remote management agent (RMA) facility. RMA provides a graphical user interface (GUI) facility for visualizing and managing JADE agents. In the main container, we can observe CARA agent (*maincaru@193.61.148.129:1099/JADE*), MSA agent (*chunker@193.61.148.129:1099/JADE*), and AAA agent (*aggregator@193.61.148.129:1099/JADE*). From the five sensor observations that have been obtained, the CARA agent has spawned various agents to monitor the ongoing activity or activities and these are launched in agent containers- i.e. Container-1 to Container-10. For instance due to the fact that the user is in the kitchen (given by the observation Mon 18-Feb-2013 14:12:40 *KitchenDoorObj SensorOn*), it can be observed that CARA launches various SARA agents to monitor kitchen-based activities e.g., *MakeTea@193.61.148.129:1099/JADE* on Container-1, *MakeSoup@193.61.148.129:1099/JADE* on Container-3., *MakePasta@193.61.148.129:1099/JADE* on Container-9, etc. Figure 10.7c shows the observation Mon 18-Feb-2013 14:15:00 *BathroomDoorObj SensorOn* has been made. At this stage, the CARA agent is shown to launch further SARA agents to monitor bathroom-based activities as shown in Fig. 10.7d, e.g. *BrushTeeth@193.61.148.129:1099/JADE* on Container-12, *Bathing@193.61.148.129:1099/JADE on Container-13*, and *WashHands@193.61.148.129:1099/JADE* on Container-14. In the meantime, Fig. 10.7c displays the result that the simple activity *MakePasta* has been identified, showing that it started at 14:12:40, when the first *observation* was made, and the current time is 14:15:14. This process proceeds as long as sensor data continues to be obtained.

10.4 A Service-Oriented SOAP-Based Smart System

The SMART system presented in Sect. 10.2 has a number of disadvantages which are outlined in Sect. 10.4.4 The main deficiency in the SMART implementation is that the system is a monolithic dotNET based software structure hosted on a single computer system, this is a traditional deployment scenario for web-accessible systems.

A more complex but flexible deployment solution would be a separation of the functionality of the SMART system into a collection of web servers that can communicate via standard web service communication technologies. This reengineering and reimplementation process is the focus of this section. A distributed collection of web services would allow system functions to be reused by a large number of smart home clients. The distributed nature also allows redundant service nodes to be deployed to provide greater system reliability. It also offers the possibility of multi-

Fig. 10.7 A snapshot of the runtime agent system

ple smart home systems to use a single collection of service nodes, thus increasing the reuse of deployed technology and increasing the overall efficiency of a system. For example, there could be five servers serving ten smart environments as opposed to one server per smart environment. The original dotNET based smart implementation uses a semantic reasoning engine called SEMWEB, to store information in a semantic ontology and allow information to be extracted by use of rules. These rules group related data by leveraging descriptions of desired patterns present in the semantic database storage structure. This SEMWEB semantic reasoning and storage mechanism is less advanced, capable and efficient than some competing solutions which have been developed in non dotNET technologies. A goal of this project is also to integrate an alternative semantic storage and reasoning system into this project by using a web service-based communication mechanism to integrate technologies that are not implemented with dotNET and are compatible with these more capable semantic tools (e.g. Java).

10.4.1 The Service-Oriented System Architecture

Figure 10.8 shows the SOA-based SMART system architecture. It uses an enterprise service bus (ESB) to provide an interface between three Java-based web service nodes and an AJAX-based web interface with communication occurring with SOAP messages. All of the service nodes, the ESB and HTML interface of the system are designed to be deployed in a single exclusive server each. Nevertheless, this distributed deployment is flexible enough that it can be reconfigured to host all of these elements on a shared single server if required, any scenario between these two deployment extremes can be catered for by modifying the ESB configuration.

The purpose of the presentation service node is to interact with other nodes in this system and submit requests to and present data returned from them in a human-readable form which is formatted with HTML. An example of the use of this service would be a web interface requesting an integrity check of the semantic ontology present in the Reasoner service. The Web interface sends a locally logged integrity check request to the presentation service. The Presentation service receives the integrity check request and then contacts the Reasoner service to perform the check. The Reasoner service performs this integrity check and returns a response which is designed to be consumed by another computer agent, the response to this request is simply a true, false or error signal. The presentation service processes these true, false or error messages to produce a meaningful message for human

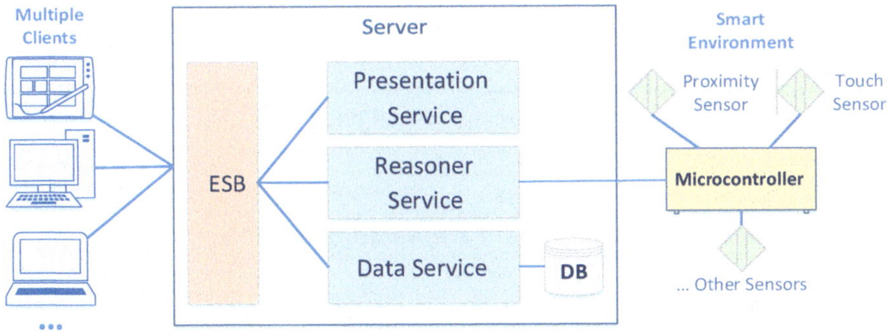

Fig. 10.8 The SOAP-based system architecture

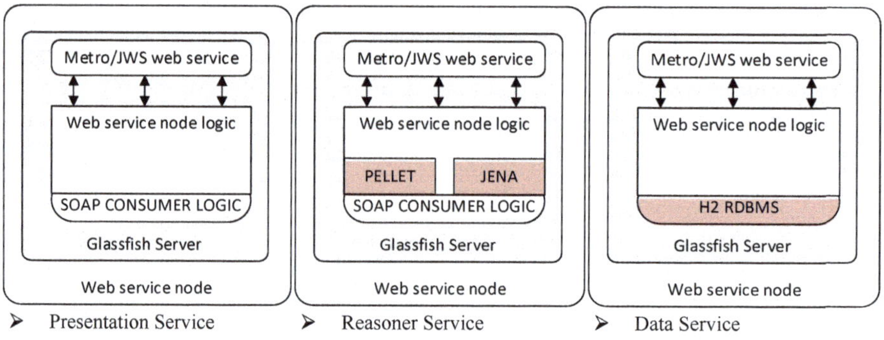

Fig. 10.9 Key services view of the ESB SOA system architecture view

consumption (e.g. "ontology valid") this message is relayed to the AJAX call the web interface made which adds this information to its local log. This Service node consumes both the Reasoner and Data Service nodes via the ESB. Figure 10.9 is a graphical representation of the Presentation service, Reasoner service and Data service components.

The Reasoner service node integrates the Semantic knowledge tools (PELLET), hosts the semantic repository (JENA) and contains the all the required supporting logic to perform all necessary SMART system operations on this semantic data (e.g. set a sensors state, determine an activity). This service is arguably the core of this reimplementation of the SMART system as it contains the most important and complex logic of any one component in this development effort. This service consumes the Data Service and interacts with the Presentation service via the ESB, the Data Service is primarily consumed to provide a persistent record of semantic reasoning operations between requests (providing state between web service requests) and log operations. The functionality of this service node is covered in detail in the technical supplement.

The Data service node provides a mechanism for this system to store, modify and access non-semantic information and importantly provides the facility to keep the state of operations between web service requests. This service consumes no other and interacts with the Presentation and Reasoner Service via the ESB. Some of the non-semantic information these stores includes: activity logs, non-semantic representation of activity preferences and system preferences.

10.4.2 The SOA Based System Implementation

This system leverages an embedded implementation of the high performance H2 relational database management system to provide a convenient and standard manner to store, manipulate and retrieve data using SQL queries. In addition to providing specialised operations to provide the functionality described above this data service also provides an interface to H2 relational database which allows any data to be stored and retrieved as uniquely keyed pairs providing a readable expandable use case for this node. For example, a future web service could need to persistent storage to record its previous operation, in that scenario a key pair with the unique name of "*furtureServicePrevOp*" could be set and retrieved from this service without any modification to its logic thus a section of a framework for future development.

The web interface is implemented using HTML, JavaScript, image files and CSS. No server-side programming technologies are required to host an instance of this interface. This reduces the complexity of this deployment. As discussed previously there the interface uses AJAX technology to send and receive SOAP requests exclusively to and from the Presentation service. This web interface has been partially complete as it only provides interfaces to functionality that was developed during this reimplementation effort. These interfaces are Smart sampler, History and Preferences.

10.4.3 The SOA Based System Interface

Figure 10.10 shows the main interface page entitled "Smart Sampler" for the SOA SMART implementation. The Smart Sampler page shows the state of the currently registered sensors and provides functionality to manually trigger sensor activations in order to check the reasoning functionality of this smart home system. In an ideal development scenario, these activations would be sent to the presentation/reasoning service directly. In addition, Smart Sampler page enables users to activate individual sensors and view recognised activities. The history view offers the ability to view and modify logs that were created when sensor interactions where set to be recorded (using "Start Recording" button on right panel). These historical records are designed to be used in future to replay a series of sensor interactions. These records are stored on the Data service which is returned to the web interface using the Presentation service via the ESB as shown in Fig. 10.11.

Figure 10.12 presents an interface to enable the user to add and update new activity preferences. The activity preferences are used for activity identification in the SMART system. For example, the following activity preference specifies parameters for the act of making green tea, which normally takes under 210s to perform and occurs at 08:20. This act involves activations of the *#ChinaCup, #GreenTea* and *#KitchenBoiler* sensors. This preference is stored in both non-semantic forms stored in the Data Service node and semantic form in the Reasoner service node. The record in the Data service is used to view preferences as this is a more efficient method than extracting the details (i.e., sensors involved, duration, and time) from the Semantic store. During update operations the original semantic preference listing is completely removed from the semantic service store and reconstructed with the necessary modifications from the data service store.

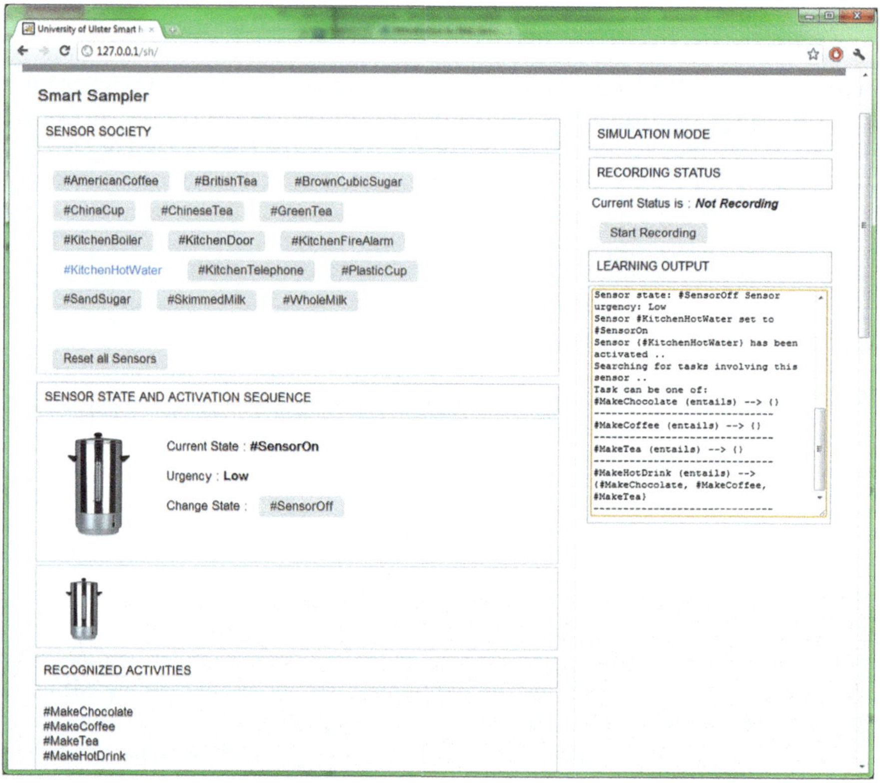

Fig. 10.10 SOA ESB Smart re-implementation UI—sensor sampler, AR and learning

10.4.4 SOA Based System Benefits and Limitations

The SOA based prototype system provides facilities to expand system resources available to a system component. In a non-distributed system, all system components are hosted in a single computer system which may not have the capacity to concurrently perform all possible operations, e.g., reasoning with a semantic knowledge tool, accessing a large data store and rendering a complex HTML interface, which may introduce a bottleneck to system performance. Notably, in the original monolithic design of this smart home system, the Reasoner service has the potential to consume 100% of CPU time on a host computer which could interfere with the operation of other processes hosted on that system, e.g., The HTML interface becomes unresponsive during semantic reasoning operation. In contrast, a distributed web service implementation such as the SOAP-based SOA system discussed in this chapter has the possibility of moving resource intensive operations to an independent dedicated computer host so that the overall system does not have any performance bottlenecks and remains responsive.

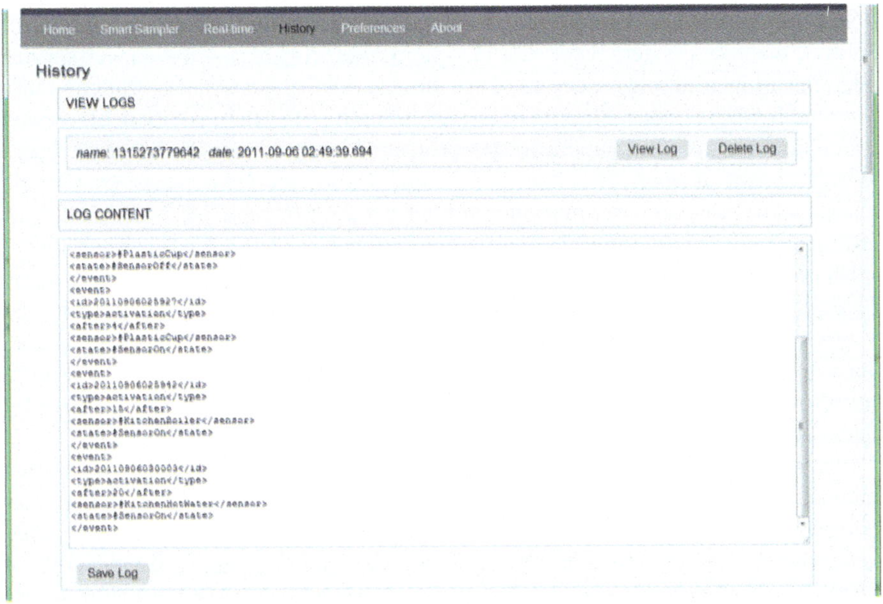

Fig. 10.11 SOA ESB Smart re-implementation UI—sensor logs

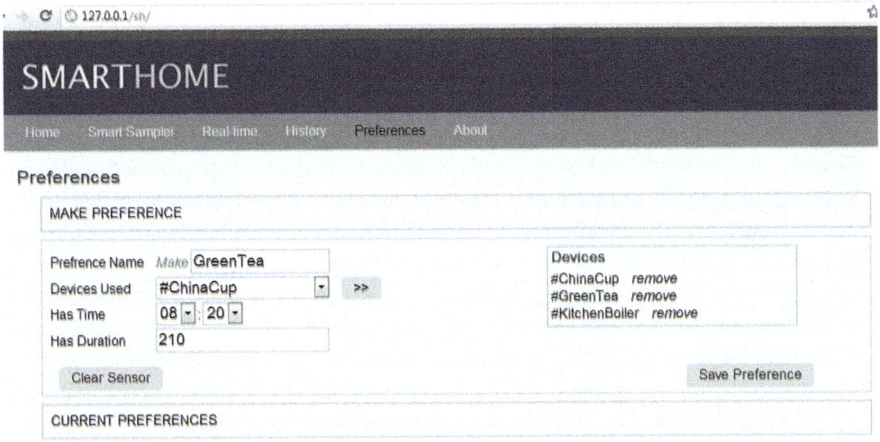

Fig. 10.12 SOA ESB Smart re-implementation UI—explicit user preferences

Secondly, the SOAP-based SOA prototype has the ability to share the capacity of deployed instances system resources. A system consisting of distributed web service nodes can allow multiple processes to use these service nodes to access functionality presented by these nodes. Sharing of the functionality of system components as web service nodes is of great benefit to this project as it introduces the scenario where the deployment of multiple instances of a smart home can be powered by a single or reduced set of computer systems without needing to replicate functionality or hardware requirements for each smart home instance.

The third advantage is that individual components can be developed in a variety of technologies. By deploying and developing this web service-based system and its supporting web service framework there is a new ability to integrate system components based on numerous differing and otherwise incompatible technologies. This allows this system to potentially use any already developed tools as part of a system component by integrating with their peers via web service messages.

The fourth advantage is the ability to independently maintain/modify individual functional due to decoupled components without affecting the whole system. As system components in a distributed system interact via web service messages they are not directly affected by changes to the internal operation and consistent code of its peer component services unless changes are made to the interfaces of an expected operation of web service functions hosted by peer web service nodes. This web service-based abstraction allows the development of web service components to be undertaken with much less coordination and compatibility testing that is required with traditional software development methods. This facilitates more independent and isolated development model for these web service nodes leading to a more efficient and less error-prone development cycle (assuming web service interfaces and responses are consistent amongst development revisions of the web service node).

Finally, the SOAP-based SOA implementation enables system functionality to be evolved in a more abstract and efficient manner. This system was developed alongside a web service development framework consisting of hardware, conceptual and software components. Production of this web service framework means that in the event that the functionality of the system needs to be expanded by the addition of new components and features there is a documented and existing ecosystem of services and tools to make this development easier. For example, in future, there may be a need to add an analytic service component to this system in order to profile usage of consumables, e.g., coffee, in each smart home environment to coordinate resupply of a geographically close set of homes to reduce costs. By use of this framework development and integration of this service could take place with only a minor amount of additional effort that is required to produce this logic in a set of Java classes.

10.5 A Multi-layered Service-Oriented REST-Based Smart System

The fourth system continues with the SOA approach but develops the web services using lightweight Representational State Transfer (REST) protocol instead of SOAP. REST-based protocol enables clients to retrieve and post information to web services without requiring a constant connection or an additional layer of information in data packets. REST-based protocol is particularly useful for sensing devices within a smart environment to post their data more efficiently. In addition, a single web service has been layered using combinations of the design patterns. In contrast to the third prototype, ESB is made redundant and communication overhead is further reduced. Another key difference is that a graph-based database (triplestore) replaces RDMS to preserve semantical data. The Apache Jena Fuseki server [24] as a triplestore was selected as it can be embedded with web services and deployed on a single server or externally on cloud servers for improved scalability and reusability. This contributes to the semantic content of linked data and subsequently supporting semantic fusion and automatic reasoning.

10.5.1 A Multi-layered SOA Based Framework

Figure 10.13 presents a multi-layered SOA based framework independent of the standalone SMART system. The framework consists of four types of actors: client devices, REST-based web service, triplestore and smart environment. The client devices running on different platforms make HTTP interactions with web services. This human-computer interface has enhanced the original SMART system by building a mobile application along with a browser-based interface. By creating the mobile application, it supports patients, carers and other stakeholders to connect with each other on the move, i.e. family members and relatives. A mobile application can further act as a sensing device utilising onboard sensors or an aggregator for other sensing devices. An Android operating system (OS) based application was developed because of its availability, popularity and large community support. However, other OS with internet access are also compatible as the data is communicated using standard messaging format. The Android application adapts simple model-view-controller (MVC) design pattern to organise the classes for visualising content from web service and make them interactive.

As can be seen in Fig. 10.13, this SOA contains five major layers: REST web service API, facade, repository, domain, and utility, which are logically separated based on the types of tasks they perform. The REST web service API exposes services to enable external client devices to consume the features/data using the HTTP asynchronously. The facade layer contains classes that perform high-level commands for complex operations by utilising multiple repository classes. It includes data access and reasoning components. The role of the data access component is to perform mul-

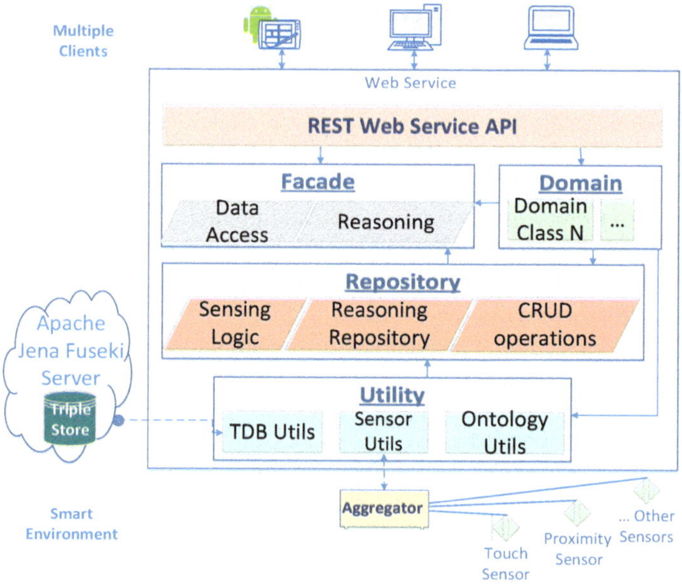

Fig. 10.13 The multilayer SOA using REST-based web service protocol

tiple operations with the triplestore in order to answer simple of complex answers. More specifically, a function in a class located in the data access's package can retrieve all unrecognised activities from the AR process and related sensors activated within a fixed period of time interval using one or more queries from the triplestore. However, the data access classes do not write queries to perform "creates, reads, updates and deletes" (CRUD) operations as this task is delegated to the repository layer. Likewise, the reasoning component makes use of reasoning functions available from the repository layer to perform activity recognition tasks. The repository layer is the core component which performs CRUD operations, parses the results retrieved from the triplestore and provides functions from the reasoner repository to enable inference using rules and variances of reasoners. The domain layer contains a set of classes that can be instantiated at run-time to temporarily store data, pass between four layers and help mapping data to and from multiple formats. Finally, the utility layer conducts low-level processes and directly communicate with triplestore via an HTTP connection, collect data from devices in the sensing environment, and support loading, manipulating and reasoning with ontology models.

Applications for assisting inhabitant's independent living in a care home can be added, such as these derived from recent inspection reports carried out by the Care Quality Commission [30]. They include managing medication dose, doctor's appointments, bedwetting assistance, and detecting inactiveness. A real-time ADL inference and simulation engine as well as the preference management and medicine dose management interfaces have been implemented to demonstrate this architecture.

10.5.2 The Multi-layered SOA Based System Implementation

The web service is central to the Android application and Apache Fuseki Server (triplestore). The Android application makes standard HTTP requests (i.e., GET, PUT, POST, and DELETE) to the web service to perform several tasks, such as CRUD operations, inference, reasoning, and other complex application-based logics. Semantic data and ontologies stored in the triplestore are retrieved and manipulated using SPARQL query language with the support of Apache Jena library. The Jersey library plays a key role in developing the RESTful web services for the function and parameter mappings of the incoming requests from the clients, as well as in producing and consuming data in various formats dynamically. In general, Jersey library is used to bind the web services with the Android application and mapping data into various object classes.

As discussed in Chap. 2 (Sect. 2.2), a diverse number of sensors and communication protocols are currently available in the market. The "*Utility*" layer in the system framework consists of packages and classes that enable extraction, storage, and processing of the data from the sensing hardware devices. In particular, the "*Sensor Utils*" component contains packages and classes that interact with third-party APIs and hardware libraries (i.e., "*.*almond*" and "*.arduino*"). Some of the key Java libraries used are WebSocket API (for Almond + router), XBee, and comPort (both for Arduino). The parallel thread classes use these classes to log the events ("*EventLogThread*"), perform device management ("*DeviceManagementThread*"), and store the data in the triple-store ("*TDBStorageThread*"). Figure 10.14 illustrates the aforementioned utility library structure.

The multi-layered SOA system uses the Securifi Almond + router to perform ambient sensing, Arduino boards for embedding sensors to objects, and Amazon Echo for voice interaction (see Fig. 10.15). The Securifi Almond + router serves as a main "*IOT Router*" because it supports multiple communication protocols, e.g. ZigBee, Z-wave and Wi-Fi, thus making it compatible with a wide range of sensor devices developed by multiple manufacturers. This makes adding and updating new sensor devices within a SH environment more easily. Similar routers supporting a wide range of sensors also include Libelium Waspmote [49], SmartThing Hub, and VeraLite. The sensors embedded within everyday objects is made possible using miniature Arduino microcontrollers with wireless (radio frequency (RF), ZigBee, or Wi-Fi) or wired (Universal Serial Bus (USB)) capabilities to transmit and collect analogue/digital sensor data; more details in [28]. One key limitation of using microcontroller-based sensing technique is that it requires expert knowledge in programming to add new sensors and configure individual components. The "Sensor *Utils*" layer enables flexible integration with other sensing devices using third-party APIs and software libraries.

The data collected from sensors are stored using classes from "*TDB Utils*" package and added to the event log queue. A class in "*TDB Utils*" performs a query and an update request in three simple steps: (1) building SPARQL query/update string, (2) using Jena classes/standard HTTP post methods to execute the request,

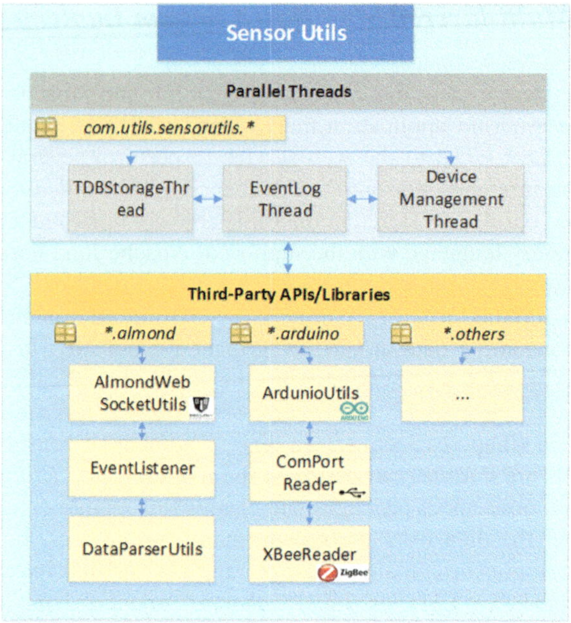

Fig. 10.14 Software: sensor utility package

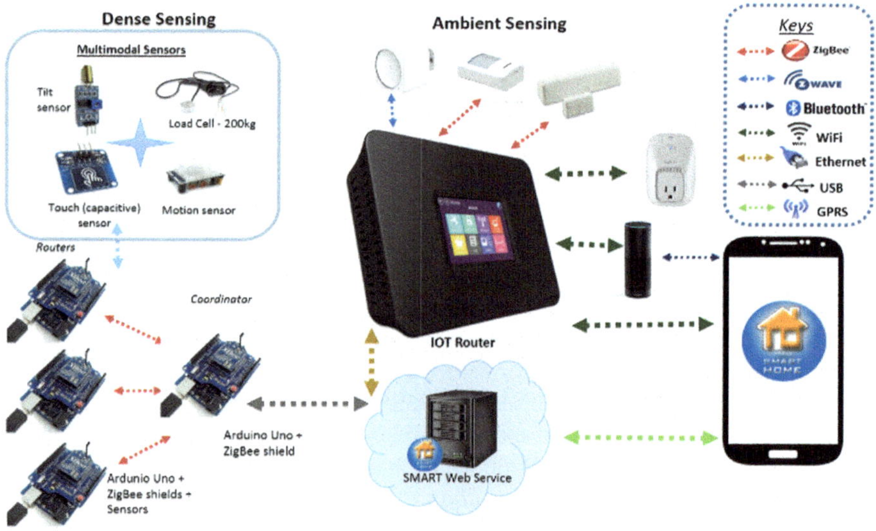

Fig. 10.15 Hardware: connectivity diagram of sensing devices

and (3) parsing the responses. The SPARQL queries are performed on the triple-store endpoint and parse the result using the *ResultSet* and *QuerySolution* classes. The standard HTTP post request can be made to perform SPARQL update using the *HttpPost*, *HttpClient*, and *HttpResponse* classes. However, the request content type is set to "*application/sparql-update*" and a static variable already defined in the Jena's *WebContent* class ("*WebContent.contentTypeSPARQLUpdate*") can be used. Furthermore, these sensor events are logged in a queue which is used by repository and façade layers to prepare, analyse and reason with the data. The "*Ontology Utils*" component in the "*Utility*" layer, provide the main support for loading ontological model, adding new assertions, and performing incremental reasoning using the Pellet reasoner.

The sensor observations and the activity reasoning results are broadcasted to the clients by the REST web service API using server-sent events (SSE) protocol. SSE protocol enables multiple clients to create a WebSocket connection results simultaneously. The web service broadcasts two SSE to the clients: one for broadcasting real-time sensor events and another with inferencing results for the clients with a session token. This sequence of events between the client device and the key components in the web service is illustrated in Fig. 10.16. As can be seen, the client Android application can listen to the sensor events in the background asynchronously by making an SSE call to "*EventBroadcaster*" function in the *SensorsCall* class located in "*SmartWebServiceAPI*" (A). To receive client-specific inferencing results, the client must obtain the session identity from the "*ReasonerCall*" first (B). The "*ReasonerCall*" is responsible for the task of listening to the sensor events from the given time, performing inferencing and then broadcast the result using "*ResultsBroadcaster*" function (B.1). Once the client receives the session token, a request can be made to "*ResultsBroadcaster*" after which the task of listening to the inferencing results associated with their session identity is initiated. Meanwhile, the client device is responsible for closing the session (C) and, if required, storing the session data separately. One key limitation of adapting SSE protocol is that all the clients will receive all the messages which can create privacy implications. An alternative messaging pattern such as publish and subscribe is also available where client subscribes to messages for a specific topic only. One such popular tool that supports publish and subscribe protocol is message queuing telemetry transport (MQTT) protocol [29].

The Android application and web browser interface can make the requests to the REST web service API using the standard HTTP protocol, currently, set to JSON format; hence the request headers need to be set appropriately. Figure 10.17 provides an example of the response messages upon executing the two *HttpGet* requests methods from the web browser for viewing the (a) notifications and (b) *Patient1*'s appointments lists. The Android application receives the same data in the JSON format and by using *org.codehaus.jackson.map.ObjectMapper* class, the data can be remapped into its respective java object instance automatically; see Jersey and Jackson libraries.

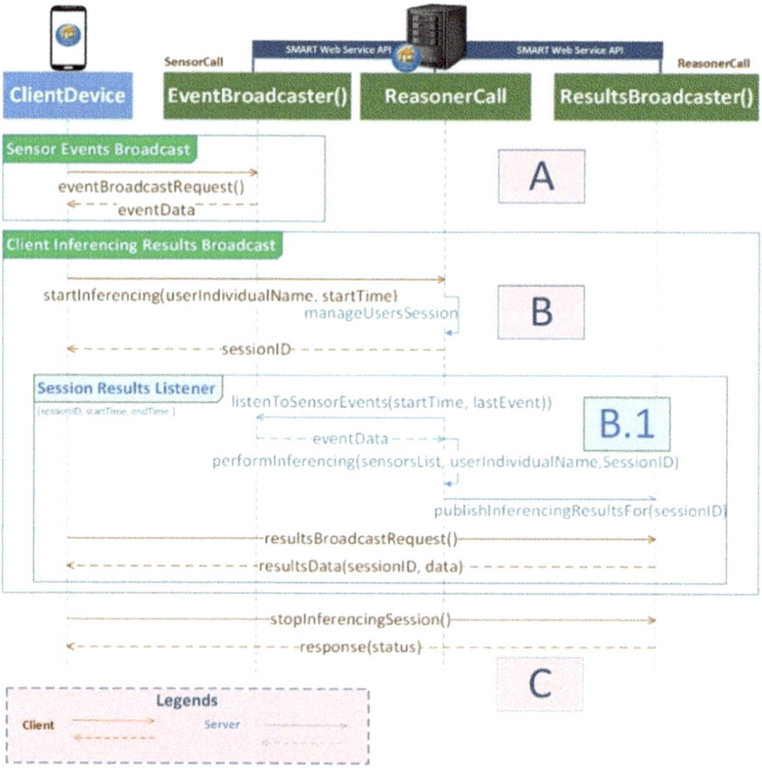

Fig. 10.16 Server-sent event (SSE) mechanism to communicate between client and web service

10.5.3 The Multi-layered SOA Based System Interface

The REST-based SOA system developed a mobile application to not only allow the inhabitant to have a better Human Computer Interface (HCI) but also act as a sensor/actuator by utilisation of the embedded sensors within the device or attach external devices using wireless connectivity (i.e. Bluetooth). The devices such as Smartwatch and Shimmer sensing devices can be used to obtain additional contextual information about the inhabit to increase accuracy in activity recognition and providing adequate assistance. Providing every patient in the care home with a smartphone may not be financially feasible. Getting the elderly to use it will pose further challenges. Therefore, providing efficient and natural HCI methods for an elderly can waiver those problems to a degree. For instance, the recent introduction of devices such as Amazon Echo that provides voice-based interaction to the system and the ability to interconnect with the smartphone and other smart devices using SmartThings can be advantageous.

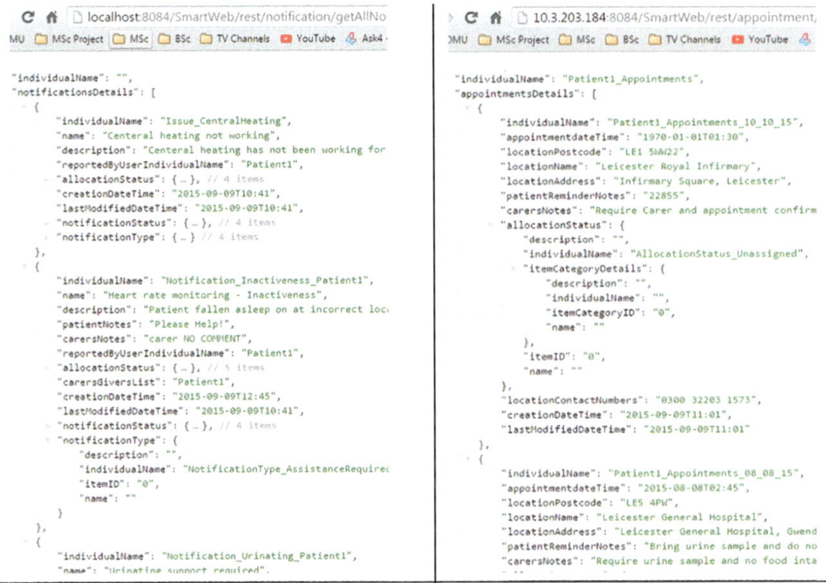

(a) Notifications list data (b) Patient1's Appointments list

Fig. 10.17 Response data from web service android application

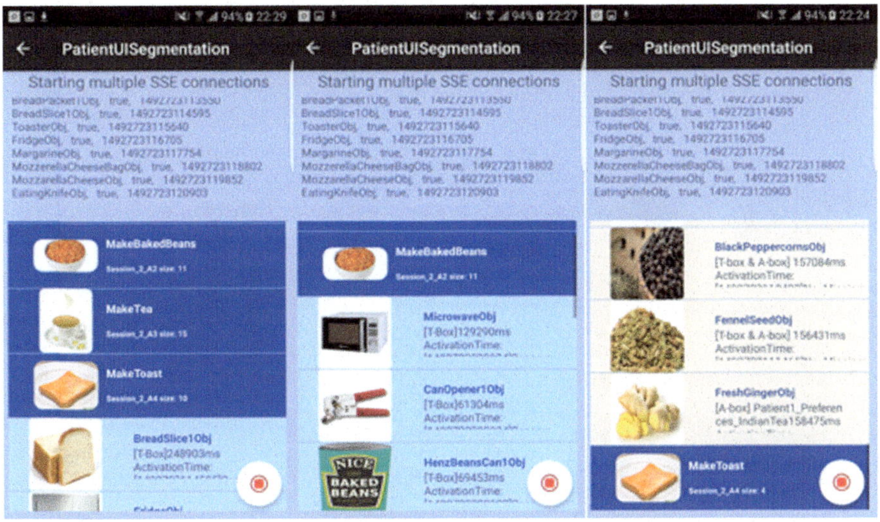

Fig. 10.18 Android application displaying semantical segmentation results three activities

The android application was developed for a smartphone with a number of features. One of which is to display the semantical segmentation results (discussed in Chap. 6) from the continuous sensor stream made available by the REST web service API. Multithreading concepts have been employed to segment each sensor events into relevant ADL threads. A single ADL thread runs the T-Box reasoning and one or more A-Box thread(s). The reasoning result and sensor events are broadcasted to the clients and the Android application continuously capture and presents the information to the inhabitant. Figure 10.18 shows a snapshot of how concurrent actions of three activities are separated into different threads and presented on the Android application. The screenshot on the left show sensor events logs on the top and an expandable list on the bottom showing *MakeToast, MakeBakedBeans*, and *MakeTea* activities. The expandable list shows respective ADL's image, thread names, and a total number of sensors segmented. As the user expand each ADL element in the expandable list, all the relevant sensors get displayed below showing sensor name, activation time, and if the action is generic or specific to the user. The partial list of segmented sensors for ADLs are illustrated at the bottom of half of all three screenshots. The *MakeTea* activity actions highlighted in yellow on the right screenshot show inhabitant's preferences being recognised during the A-box reasoning process.

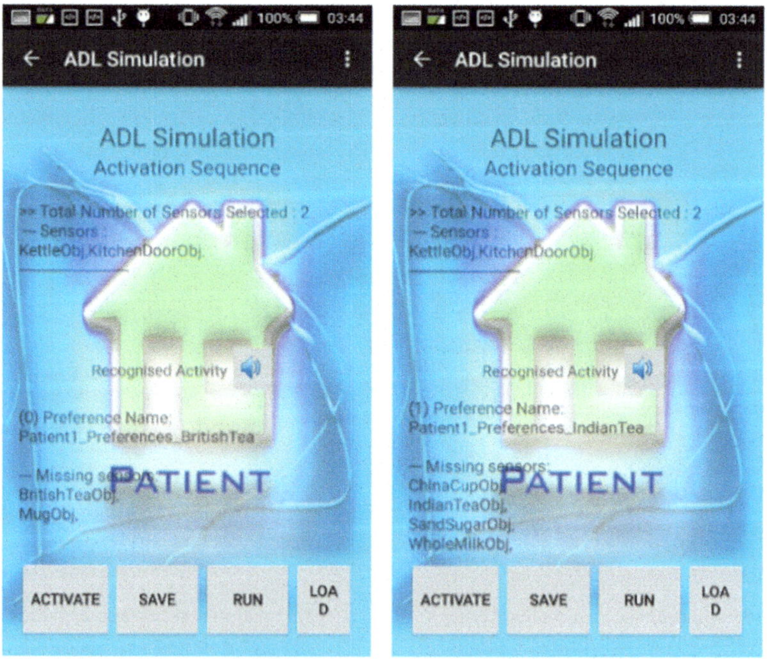

(a) Simulation result (1) (b) Simulation result (2)

Fig. 10.19 ADL simulation environment for ADL preferences matching

Another key feature developed was the simulation interface to load a set of actions for one or more activities and perform SPARQL based activity inferencing. The requests are made from the Android application to perform the simulation and reasoning tasks on the web service. The activation of simulated sensors and results are displayed in Fig. 10.19. The Text-to-Speech feature is also used to utter the resultant output (using speaker icon). The activity recognition algorithm is performed by the web service using a knowledge-driven approach. Currently, only pre-defined user ADL preferences are used to match against the activated sensors. The aim of the matching process is to find the related user preference(s) and other inactivated sensor object(s) from those individual preference(s) to complete the activity. The user can create and manage their preferences using the interactive forms depicted in Fig. 10.20. Moreover, a user can browse through all the sensors by location and type, add new sensors or update individual sensors using the interface in Fig. 10.20b.

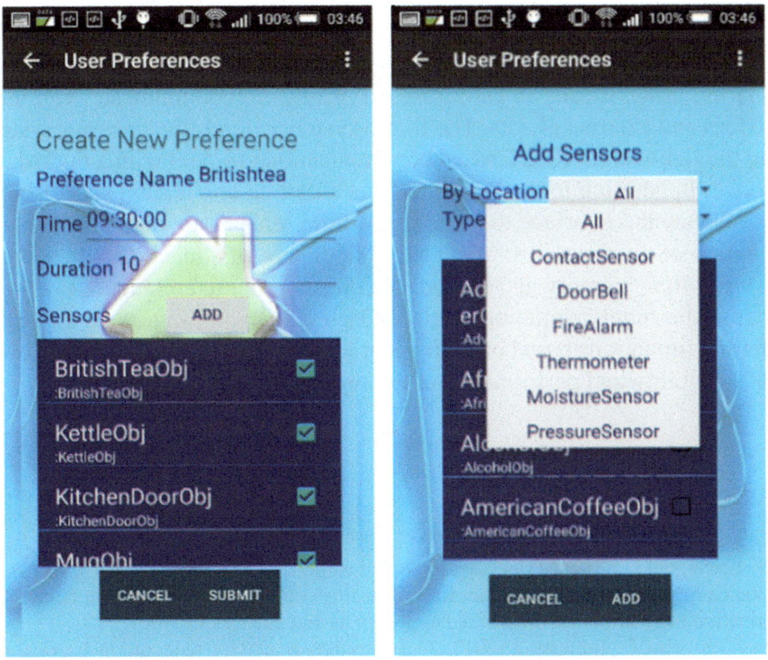

(a) Creating a new preference

(b) Adding sensors to the preference

Fig. 10.20 User preference management interface

10.6 Summary

This chapter presents four prototype systems which were developed for testing and evaluating various activity recognition approaches investigated in previous chapters. These prototype systems are categorised based on their architecture styles into standalone, multi-agent and SOA with SOAP protocol and multi-layered SOA with REST protocol. This reflects and closely corresponds to the evolution of the latest technologies in software engineering and smart cyber-physical systems. The initial standalone SMART system was feature rich, simple and efficient to recognise single-user sequential activities with user preferences. Nevertheless, the nature of standalone restricted the ability for the system to be reused by other systems, add new features in unmanageable monolithic coding style and components with licencing cost. The second prototype system introduced agent-based system architecture with a Java-based application to recognise activities. The latter two systems adopted a distributed architecture with open source components to make the system more accessible for other platforms over the internet connection. The third system implemented a SOAP-based service-oriented architecture with multiple web services collaboratively sharing tasks. It uses an enterprise service bus to communicate with internal web services and external clients. The fourth system extends the previous system by adopting the lightweight REST-based communication protocol with a multi-layered web service. In addition to the web interface, a mobile application was also developed to enhance the interface to receive real-time sensor data and AR results. The key limitation of the fourth prototype is the single use of a machine which requires very high run-time memory and number of cores in order to conduct AR tasks. This chapter describes implementation details of each system prototype, and discuss their strengths and limitations based on their accuracy and performance. Future work will focus on scalability and performance of the third and fourth systems.

References

1. Chen L, Nugent C, Al-Bashrawi A (2009) Semantic data management for situation-aware assistance in ambient assisted living. In: Proceedings of the 11th international conference on information integration and web-based applications and services - iiWAS '09
2. Chen L, Nugent C, Rafferty J (2013) Ontology-based activity recognition framework and services. In: Proceedings of international conference on information integration and web-based applications and services - IIWAS '13, pp 463–469
3. Wang X, Wang J, Wang X, Chen X (2017) Energy and delay tradeoff for application offloading in mobile cloud computing. IEEE Syst J 11:858–867
4. Semantic Web: Semantic Web/RDF Liberary for C#.NET. http://semanticweb.org/wiki/SemWeb-DotNet.html
5. Della Valle E, Grossniklaus M(2010) C-SPARQL: a continuous query language for rdf data streams. Int J Semant Comput 04:3–25
6. W3C: Euler proof mechanism. http://www.agfa.com/w3c/euler/
7. Nugent CD, Mulvenna MD, Hong X, Devlin S (2009) Experiences in the development of a smart lab. Int J Biomed Eng Technol 2:319–331

8. Löhr KP (2003) Automatic mediation between incompatible component interaction styles. In: Proceedings of the 36th annual HAWAII international conference on system sciences, HICSS 2003
9. W3C: W3C semantic web knowledge base, Listing of tools by programming. https://www.w3.org/2001/sw/wiki/Category:Programming_Language
10. Apple: Objective- C programming Documentation. https://developer.apple.com/library/archive/documentation/Cocoa/Conceptual/ObjectiveC/Introduction/introObjectiveC.html
11. Perl: Perl Programming Documentation. http://perldoc.perl.org/
12. Abu-Eid V (2008) Raising web service updates efficiency with dynamic technologies. In: Proceedings - international conference on next generation web services practices, NWeSP 2008
13. Microsoft: Microsoft dotNET technology information website. http://www.microsoft.com/net/
14. Jensen PA (1969) The design of multiple-line redundant networks. IEEE Trans Reliab R-18:39–44
15. Wooldridge M, Jennings NR (1995) Intelligent agents: theory and practice. Knowl Eng Rev 10(2):115-52
16. Srivastava SK (1999) Applications of Intelligent agents. Electron Inf Plan
17. Jennings NR, Wooldridge MJ (1998) Agent technology: foundations, applications and markets.
18. Jennings NR (2002) An agent-based approach for building complex software systems. Commun ACM
19. Bellifemine F, Poggi A, Rimassa G (2001) JADE: a FIPA2000 compliant agent development environment. In: International conference on autonomous agents and multiagent systems
20. Horrocks I, Patel-Schneider PF, Bechhofer S, Tsarkov D (2005) OWL rules: a proposal and prototype implementation. Web Semant
21. Khan JA, Kumar S (2015) OWL, RDF, RDFS inference derivation using Jena semantic framework and pellet reasoner. In: 2014 International conference on advanced engineering and technology research. ICAETR 2014. 0–7
22. Friedman-Hill E (2008) Jess the rule engine for java platform
23. Mei J, Bontas EP Technical Reports: reasoning paradigms for owl ontologies. http://www.ag-nbi.de/research/owltrans/
24. Apache Apache Jena. https://jena.apache.org/
25. Science C, Lanka S (2013) Application of design pattern in the JDBC programming. In: 2013 8th international conference on computer science & education, pp 1037–1040
26. Ali M, Elish MO (2013) A comparative literature survey of design patterns impact on software quality. International conference on information science and applications (ICISA), pp 1–7
27. Zhang C, Budgen D, Drummond S (2012) Using a follow-on survey to investigate why use of the visitor, singleton & facade patterns is controversial. In: Proceedings of the ACM-IEEE international symposium on empirical software engineering and measurement - ESEM '12, p 79
28. Triboan D, Chen L, Chen F, Wang Z (2016) Towards a service-oriented architecture for a mobile assistive system with real-time environmental sensing. TSINGHUA Sci Technol 21:581–597
29. MQTT: Message queuing telemetry transport (MQTT). http://mqtt.org/
30. Care Quality Commission: Care Quality Commission. https://www.cqc.org.uk/about-us
31. Abburu S (2012) A survey on ontology reasoners and comparison. Int J Comput Appl 57:33–39
32. Jersey: Chapter 15. Server-Sent Events (SSE) Support. https://jersey.java.net/documentation/latest/sse.html#d0e11582

Index

L. Chen and C. D. Nugent, *Human Activity Recognition and Behaviour Analysis*, https://doi.org/10.1007/978-3-030-19408-6

Printed by Printforce, the Netherlands